Annals of Mathematics Studies

Number 101

RANDOM FOURIER SERIES WITH APPLICATIONS TO HARMONIC ANALYSIS

BY

MICHAEL B. MARCUS

AND

GILLES PISIER

PRINCETON UNIVERSITY PRESS

AND

UNIVERSITY OF TOKYO PRESS

PRINCETON, NEW JERSEY

1981

The publishers are grateful for
the assistance of the Andrew W. Mellon Foundation
in the publication of this book

Published in Japan exclusively by
University of Tokyo Press;
In other parts of the world by
Princeton University Press

Printed in the United States of America
by Princeton University Press, Princeton, New Jersey

Library of Congress Cataloging in Publication data will
be found on the last printed page of this book

CONTENTS

Random Fourier Series With

Applications to Harmonic Analysis

CHAPTER I
INTRODUCTION

Abstract: Necessary and sufficient conditions are obtained for the a.s. uniform convergence of random Fourier series on locally compact Abelian groups and on compact non Abelian groups. Many related results such as a central limit theorem are obtained. The methods developed are used to study questions in harmonic analysis which are not intrinsically random.

In a series of three papers published in 1930 and 1931, Paley and Zygmund [44] studied a variety of problems concerning series of independent random functions and raised the question of the uniform convergence a.s. of the random Fourier series

(1.1) $$\sum_{n=0}^{\infty} c_n \varepsilon_n e^{inx}, \quad x \in [0, 2\pi],$$

where $\{c_n\}$ are real numbers, $\sum_{n=0}^{\infty} c_n^2 = 1$, and $\{\varepsilon_n\}$ is a Rademacher sequence, i.e. a sequence of independent random variables each one taking the value plus and minus one with equal probability. They also considered (1.1) with $\{\varepsilon_n\}$ replaced by a Steinhaus sequence $\{e^{i2\pi\omega_n}\}$, where $\{\omega_n\}$ are independent random variables each one uniformly distributed on $[0,1]$. Their results were that the series (1.1) converges uniformly a.s. if

(1.2) $$\sum_{n=2}^{\infty} c_n^2 (\log n)^{1+\varepsilon} < \infty, \quad \varepsilon > 0$$

but not necessarily if $\varepsilon = 0$. Also they introduced the numbers

$$s_j = \left(\sum_{n=2^j}^{2^{j+1}-1} c_n^2 \right)^{1/2}$$

and showed that

$$\sum_{j=1}^{\infty} s_j = \infty \tag{1.3}$$

implies that for almost all $x \in [0, 2\pi]$, (1.1) is unbounded a.s. (Their results are stated for both the Rademacher and Steinhaus series but Kahane [26], pp. 177, points out that the implication of (1.3) is proved only for the Steinhaus series.) It is remarkable that these early results are so sharp. Under certain smoothness conditions on the $\{c_n\}$ they are close to being necessary and sufficient.

In 1954 Salem and Zygmund [55] returned to this problem. They slightly modified (1.1) to consider series of the form

$$\sum_{n=0}^{\infty} c_n \varepsilon_n \cos(nx + \alpha_n), \quad x \in [0, 2\pi], \tag{1.4}$$

where $0 \leq \alpha_n \leq 2\pi$ are real numbers and also considered the case with $\{e^{i2\pi\omega_n}\}$ replacing $\{\varepsilon_n\}$. They sharpened (1.2) obtaining

$$\sum_{j=2}^{\infty} \frac{\left(\sum_{n=j}^{\infty} c_n^2 \right)^{1/2}}{j(\log j)^{1/2}} < \infty \tag{1.5}$$

as a sufficient condition for the uniform convergence a.s. of (1.4). A noteworthy aspect of this result is that together with (1.3) it shows that for s_j non-increasing, $\Sigma s_j < \infty$ is a necessary and sufficient condition for the uniform convergence a.s. of (1.4).

In 1951 Hunt [20] considered (1.1) but with $\{\varepsilon_n\}$ replaced by rather arbitrary sequences of independent mean zero random variables including those without second moments. However, it was Kahane who introduced the methodology by which generalizations of (1.1) and (1.4) should be studied. Following Kahane [26] let

$$X(x) = \sum_{n=0}^{\infty} c_n \eta_n \cos(nx + \Phi_n), \quad x \in [0, 2\pi], \tag{1.6}$$

where $\{\eta_n e^{i\Phi_n}\}$ is a sequence of symmetric complex valued random variables (η_n and Φ_n real) with $E|\eta_n|^2 = 1$. By the three series theorem this series converges a.s. for each $x \in [0, 2\pi]$. It is in this sense that the stochastic process $X(x)$ is defined. We introduce a Rademacher sequence $\{\varepsilon_n\}$ independent of $\{\eta_n e^{i\Phi_n}\}$. The series

$$\sum_{n=0}^{\infty} c_n \varepsilon_n \eta_n \cos(nx + \Phi_n), \quad x \in [0, 2\pi] \tag{1.7}$$

and (1.6) are stochastically equivalent. Let $(\Omega_1, \mathcal{F}_1, P_1)$ be the probability space for $\{\eta_n e^{i\Phi_n}\}$ and $(\Omega_2, \mathcal{F}_2, P_2)$ the probability space for $\{\varepsilon_n\}$. Then (1.7) is defined on the product space $(\Omega_1 \times \Omega_2, \mathcal{F}_1 \times \mathcal{F}_2, P_1 \times P_2)$. On this space, for $\omega_1 \in \Omega_1$, consider

$$X(x, \omega_1) = \sum_{n=0}^{\infty} c_n \varepsilon_n \eta_n(\omega_1) \cos(nx + \Phi_n(\omega_1)), \quad x \in [0, 2\pi]. \tag{1.8}$$

This is a Rademacher series of the type (1.4). Therefore the question of the uniform convergence a.s. of (1.6) reduces to one of the uniform convergence a.s. of a family of Rademacher series of the form (1.4). To demonstrate the usefulness of this approach we show that (1.5) is also a sufficient condition for the uniform convergence a.s. of (1.6). Using the sufficient condition of Salem and Zygmund we have that (1.8) converges uniformly a.s. if

$$(1.9) \qquad \sum_{j=2}^{\infty} \frac{\left(\sum_{n=j}^{\infty} c_n^2 \eta_n^2(\omega_1)\right)^{1/2}}{j(\log j)^{1/2}} < \infty .$$

By Jensen's inequality (1.5) implies (1.9) for $\omega_1 \subset \bar{\Omega}_1 \subset \Omega_1$ with $P(\bar{\Omega}_1) = 1$. Therefore (1.8) converges uniformly a.s. on $(\Omega_2, \mathcal{F}_2, P_2)$ for a set of measure one with respect to P_1. It follows by Fubini's theorem that (1.6) converges uniformly a.s. This argument demonstrates the paramount importance of the Rademacher series. In [35], concentrating on the Rademacher series and using Kahane's contraction principle as generalized by Hoffmann-Jørgensen it was shown that for $\{c_n\}$ non-increasing (1.5) is necessary and sufficient for the uniform convergence of (1.6). Nevertheless, it was already known ([55]) that (1.5) is not necessary in general.

Let $\{g_n\}$ be a sequence of independent normal random variables with mean zero and variance 1, $(N(0,1))$. Also we will denote by $\{\tilde{g}_n\}$ independent normalized complex valued normal random variables, i.e. $\tilde{g}_n = g_n/\sqrt{2} + i g_n'/\sqrt{2}$ where $\{g_n\}$ is given above and $\{g_n'\}$ is an independent copy of $\{g_n\}$. If we replace $\{\varepsilon_n\}$ by $\{\tilde{g}_n\}$ in (1.1) the series represents a stationary complex valued Gaussian process.

Beginning in the early sixties there was a surge of interest in the study of sample path properties of Gaussian processes. Dudley had the idea of using properties of the metric entropy of a natural norm associated with these processes as a criterium for determining whether the processes have continuous sample paths a.s. In [6] he obtained a sufficient condition for continuity, which when applied to random Fourier series which were also Gaussian processes, was strictly stronger than (1.5) and which, in the case of stationary Gaussian processes was possibly also a necessary condition. This motivated the work in [34], [23] and [36] which considered random Fourier series in the context of real valued stationary Gaussian processes, i.e. the series

$$\sum_{n=0}^{\infty} c_n[g_n \cos nx + g_n' \sin nx], \quad x \in [0, 2\pi] .$$

In 1974 Fernique [8] solved the problem of finding necessary and sufficient conditions for the continuity of stationary Gaussian processes. This, and the results of [36], are the crucial ingredients in our work and enabled us in [39] to obtain necessary and sufficient conditions for the uniform convergence a.s. of the series (1.6). One interesting aspect of our result is that with no additional difficulty the integers in the trigonometric functions can be replaced by real numbers. In fact, it turns out that we may even replace R or Z by any locally compact Abelian group and we may treat the general case of random Fourier series on locally compact Abelian groups. A statement of these results and an outline of the proofs was given in [40]. In this book we present complete proofs and also consider random Fourier series on compact non-Abelian groups.

Let G be a locally compact Abelian group with identity element 0. Let $K \subset G$ be a compact symmetric neighborhood of 0. Let Γ denote the characters of G and let $A \subset \Gamma$ be countable. Since we only consider Fourier series with spectrum in A, and since A generates a closed separable subgroup of Γ, in all that follows, we will assume (without restricting the generality) that Γ itself is separable, so that the compact subsets of G are metrizable. We define the following sequences of random variables indexed by $\gamma \in A$: $\{\varepsilon_\gamma\}$ a Rademacher sequence, $\{g_\gamma\}$ independent $N(0,1)$ random variables and $\{\xi_\gamma\}$ complex valued random variables (not necessarily independent) satisfying

$$\sup_{\gamma \in A} E|\xi_\gamma|^2 < \infty \quad \text{and} \quad \inf E|\xi_\gamma| > 0 . \tag{1.10}$$

Let $\{a_\gamma\}$ be complex numbers satisfying $\sum_{\gamma \in A} |a_\gamma|^2 < \infty$ and consider the random Fourier series

$$Z(x) = \sum_{\gamma \in A} a_\gamma \varepsilon_\gamma \xi_\gamma \gamma(x), \quad x \in K \tag{1.11}$$

where the two sequences $\{\varepsilon_\gamma\}$ and $\{\xi_\gamma\}$ are independent. For each fixed $x \in K$ the series converges a.s. so the sum is well defined. In Theorem 1.1 we give a necessary and sufficient condition for the series (1.11) to be uniformly convergent a.s. on K. Note that if the $\{\xi_\gamma\}$ are independent and symmetric then the $\{\varepsilon_\gamma\}$ are superfluous in (1.11).

Define $K \oplus K = \{x+y \mid x \in K, y \in K\}$ and in a similar fashion define $\underset{n}{\oplus} K = \{x_1 + \cdots + x_n \mid x_1 \in K, \cdots, x_n \in K\}$. Let μ denote the Haar measure on G and let $\mu_n = \mu(\underset{n}{\oplus} K)$. We define a translation invariant pseudo-metric d on G by

$$(1.12) \qquad d(x,y) = \sigma(x-y) = \left(\sum_{\gamma \in A} |a_\gamma|^2 |\gamma(x) - \gamma(y)|^2\right)^{1/2}$$
$$= \left(\sum_{\gamma \in A} |a_\gamma|^2 |\gamma(x-y) - 1|^2\right)^{1/2}.$$

(To see the motivation for this note that when $E|\xi_\gamma|^2 = 1$ for all $\gamma \in A$ then $\sigma(x-y) = (E|Z(x) - Z(y)|^2)^{1/2}$.) Note that $\sigma(x)$ is a non-negative function on $K \oplus K$. Define

$$(1.13) \qquad m_\sigma(\varepsilon) = \mu(x \in K \oplus K \mid \sigma(x) < \varepsilon)$$

and

$$\overline{\sigma(u)} = \sup\{y \mid m_\sigma(y) < u\}.$$

Since $0 \le m_\sigma(\varepsilon) \le \mu_2$ the domain of $\bar{\sigma}$ is the interval $[0, \mu_2]$. We see that $\overline{\sigma(u)}$ viewed as a random variable on $[0, \mu_2]$ has the same probability distribution with respect to normalized Lebesgue measure on $[0, \mu_2]$ that $\sigma(x)$ has with respect to normalized Haar measure on $K \oplus K$. In keeping with classical terminology we call $\bar{\sigma}$ the non-decreasing rearrangement of σ (with respect to $K \oplus K$). In terms of μ, σ and K we define the integral

$$(1.14) \qquad I(K,\sigma) = I(\sigma)$$
$$= \int_0^{\mu_2} \frac{\overline{\sigma(s)}}{s\left(\log \frac{4\mu_4}{s}\right)^{1/2}}\, ds.$$

We can now state our main result. (The first assertion was proved in [36] for $G = \mathbf{R}$.)

THEOREM 1.1. *Employing the notation and definitions given above let* $\|Z\| = \sup_{x \in K} |Z(x)|$. *If* $I(\sigma) < \infty$ *the series (1.11) converges uniformly a.s. and*

$$(1.15) \qquad (E\|Z\|^2)^{1/2} \le C\,(\sup_{\gamma \in A} E|\xi_\gamma|^2)^{1/2}\left[\left(\sum_{\gamma \in A} |a_\gamma|^2\right)^{1/2} + I(\sigma)\right]$$

where C *is a constant independent of* $\{a_\gamma\}$ *and* $\{\gamma | \gamma \in A\}$. *Conversely let* $\{\gamma_k, k = 1, 2, \cdots\}$ *be an ordering of* $\gamma \in A$ *and let* $\{a_k\}$, $\{\varepsilon_k\}$ *and* $\{\xi_k\}$ *be the corresponding ordering of* $\{a_\gamma\}$, $\{\varepsilon_\gamma\}$ *and* $\{\xi_\gamma\}$. *If* $I(\sigma) = \infty$ *then for all open sets* $U \subset K$

$$(1.16) \qquad \sup_n \sup_{x \in U} \left|\sum_{k=0}^{n} a_k \varepsilon_k \xi_k \gamma_k(x)\right| = \infty$$

on a set of measure greater than zero. (Note that neither (1.10) nor (1.12) depend on the order of $\{\gamma_k\}$ *so that the implications of* $I(\sigma) < \infty$ *are also valid for all orderings* $\{\gamma_k\}$ *of* $\gamma \in A$.)

If the $\{\xi_k\}$ *are independent the event in (1.16) occurs with probability one but we will show that in the generality considered here we can only say that it occurs on a set of measure greater than zero.*

Theorem 1.1 states that the condition $I(\sigma) < \infty$ is necessary and sufficient for the a.s. convergence in $C(K)$ of the series (1.11). In fact, this condition also characterizes the series (1.11) such that the process $\{Z(x)\}_{x \in K}$ has a version with continuous sample paths (cf. Remark 2.4.4 below). Moreover, if G is a countable union of compact subsets, so that $C(G)$ is a Fréchet space, then we may state: $I(\sigma) < \infty$ if and only if the random series (1.11) is a.s. the Fourier series of a continuous function on G. Indeed, the series (1.11) is a.s. the Fourier series of a continuous function on G if and only if it converges a.s. in $C(G)$ (cf. Remark 2.4.4 below which has an obvious extension for separable Fréchet spaces).

Although symmetry plays an important role in Theorem 1.1, it is not essential when the variables $\{\xi_\gamma\}$ are independent:

COROLLARY 1.2. *Employing the notation and definitions given above assume that* $\{\xi_\gamma\}$, *in addition to satisfying (1.10) are also independent and satisfy* $E\,\xi_\gamma = 0$ *for all* $\gamma \in A$. *Then all the results of Theorem 1.1 hold with* $Z(x)$ *replaced by* $\tilde{Z}(x) = \sum_{\gamma \in A} a_\gamma \xi_\gamma \gamma(x)$, $x \in K$ *and* $\sum_{k=0}^{n} a_k \varepsilon_k \xi_k \gamma_k(x)$ *in (1.16) replaced by* $\sum_{k=0}^{n} a_k \xi_k \gamma_k(x)$. *(Also* C *must be replaced by 4C in (1.15).)*

We also prove in Chapter III, Theorem 1.4 an analogue of (1.15) for increments $Z(x) - Z(y)$, $x, y \in K$.

When $G = R$, the real line, Theorem 1.1 doesn't completely answer the question considered in [26]. Let

$$Q(x) = \sum_{k=0}^{\infty} a_k \varepsilon_k \rho_k \cos(\lambda_k t + \Phi_k), \quad t \in [0,1] \tag{1.17}$$

where $\{a_k\} \in \ell^2$ are real, $\{\varepsilon_k\}$ is a Rademacher sequence, $\{\lambda_k\}$ are real numbers and $\{\rho_k e^{i\Phi_k}\}$, $(\rho_k, \Phi_k$ real), are random variables satisfying $\sup E\,\rho_k^2 < \infty$ and $\liminf_{k\to\infty} E|\rho_k| > 0$. We have the following version of Theorem 1.1 in this case.

THEOREM 1.3. *Let* $\|Q\| = \sup_{t \in [0,1]} |Q(t)|$ *and let* σ *be defined as in (1.12) then if* $I(\sigma) < \infty$ *the series (1.17) converges uniformly a.s. and*

$$(E\|Q\|^2)^{1/2} \le C'(\sup_k E\,\rho_k^2)^{1/2}\left[\left(\sum_{k=0}^{\infty} a_k^2\right)^{1/2} + I(\sigma)\right] \tag{1.18}$$

where C' *is a constant independent of* $\{a_k\}$. *If* $I(\sigma) = \infty$ *then for all open sets* $U \subset [0,1]$

(1.19) $$\sup_n \sup_{t \in U} \left|\sum_{k=0}^{n} a_k \varepsilon_k \rho_k \cos(\lambda_k t + \Phi_k)\right| = \infty \text{ a.s.}$$

on a set of probability greater than zero.

Note that this result is an immediate consequence of Theorem 1.1 when $I(\sigma) < \infty$. However when $I(\sigma) = \infty$ we have to show that

(1.20) $$\sup_n \sup_{t \in U} \left|\sum_{k=0}^{n} a_k \varepsilon_k \rho_k e^{i\Phi_k} e^{i\lambda_k t}\right| = \infty \text{ a.s.}$$

on a set of positive probability implies (1.19). This is done in Chapter III §2.

We consider one more variation of Theorem 1.1. The proof of Theorem 1.1 uses an argument similar to the closed graph theorem (cf. the proof of Corollary 2.4.15). Consequently certain constants that enter into estimates of $E\|Z\|^2$ cannot be given numerical values. When G is compact and $K = G$ we can obtain a two-sided bound for $E\|Z\|$ which can be computed.

THEOREM 1.4. *Consider the hypothesis of Theorem 1.1 with the additional condition that* G *is compact and* $K = G$. *Then*

(1.21) $$C_1(\inf_{\gamma \in A} E|\xi_\gamma|)\left[\left(\sum_{\gamma \in A} |a_\gamma|^2\right)^{1/2} + I(\sigma)\right] \le E\|Z\|$$

$$\le C_2(\sup_{\gamma \in A} E|\xi_\gamma|^2)^{1/2}\left[\left(\sum_{\gamma \in A} |a_\gamma|^2\right)^{1/2} + I(\sigma)\right]$$

for numerical constants $0 < C_1, C_2 < \infty$.

The proof of Theorem 1.4 is given in Chapter III §3. In general Chapter III is devoted to the continuity properties of random Fourier series on locally compact Abelian groups.

Chapter II contains preliminaries about metric entropy of sets on which there is a translation invariant metric and the relation between metric entropy and the non-decreasing rearrangement of the metric. It also contains a discussion of the continuity of Gaussian processes adapted to our needs and gives some of the fundamental inequalities involving sums of independent Banach space valued random variables.

In Chapter IV, we show that the random Fourier series (1.11), when considered as a $C(K)$-valued random variable, verifies the central limit theorem and the law of the iterated logarithm as soon as it is a.s. continuous and $\sup_k E|\xi_k|^2 < \infty$. In fact, this is essentially a corollary of Chapter III and the results of [36]; we indicate briefly a direct argument, simpler than the one in [36].

We also introduce the space of a.s. continuous Rademacher random Fourier series, which can be equipped with a natural Banach space structure. Curiously, this Banach space turns out to be of cotype 2, a fact which is closely related to the results of Chapters III and IV.

In Chapter V, we turn to Random Fourier series on compact non-Abelian groups. In §5.3 we extend practically all the results of Chapter III to the non-commutative setting. This extension requires a number of preliminary results on random matrices which are given in §5.1 and §5.2. We have presented these preliminaries in detail, since the ideas involved seem to give a simpler approach to several problems studied in the literature (see, for example, Remark 6.2.7).

We also obtain a generalization of Kahane's theorem on the integrability of Rademacher series with coefficients in a Banach space with Kwapien's improvement [28]. (In the scalar case, the non-commutative extension of Khintchine's inequality is due to Figa-Talamanca and Rider, cf. [11], [13].) Let $\{d_i | i \in N\}$ be a sequence of integers. Denote by $U(n)$ the compact group of all $n \times n$ unitary matrices. Consider $\Omega = \prod_{i \in N} U(d_i)$ and let $\mathbf{P}$ be the normalized Haar measure on Ω. For $\omega \in \Omega$, we denote $\{u^i_{jk}(\omega)\}_{i \le j,k \le d_i}$ the matrix which is the i-th component of ω. Let

$\{x^i_{jk} | 1 \le j, k \le d_i\}$ be elements in an arbitrary Banach space. Define

$$Y_i(\omega) = \sum_{1 \le j,k \le d_i} u^i_{jk}(\omega) x^i_{jk} .$$

Kahane's theorem is extended as follows: If the series $S = \sum_{i=1}^{\infty} Y_i$ converges a.s. in norm, then necessarily $E \exp \varepsilon \|S\|^2 < \infty$ for some $\varepsilon > 0$. (In fact, for all $\varepsilon > 0$, for the same simple reason as in [28].) Our proof uses the same basic idea as [28].

In Chapter VI we present several applications of random Fourier series to Harmonic Analysis due to the second named author (cf. [47], [48]). These results form a natural continuation of the preceding ones. Let G be a compact group. In Section 6.1 we consider again the space formed by the functions f in $L_2(G)$ such that the randomized Fourier series of f is a.s. continuous. The corresponding Banach space is denoted $C_{a.s.}(G)$. We show (cf. Theorem 6.1.1) that $C_{a.s.}(G)$ can be identified with the predual of a space of Fourier-multipliers. This gives a new characterization, this time Fourier theoretical, of a.s. continuous random Fourier series. To be more precise, a square integrable function f belongs to $C_{a.s.}(G)$ if and only if f has an expansion

$$f = \sum_{n=1}^{\infty} h_n * k_n$$

with

$$\sum_{n=1}^{\infty} \|h_n\|_2 \|k_n\|_\phi < \infty ,$$

where $\| \ \|_\phi$ is the norm in the Orlicz space $L_\phi(G)$ with $\phi(t) = t(1+\mathrm{Log}(1+t))^{1/2}$ for all $t \in \mathbf{R}_+$. (The space L_ϕ is often referred to as the class $L\sqrt{\mathrm{Log}\, L}$.) As a corollary of this we show in §6.2 that (even in the non-Abelian case) Sidon sets are characterized by the order of growth of their $\Lambda(p)$-constants, thus answering a question raised by Rudin in [53] (see Theorem 6.2.3). This application was obtained first in [48], and it suggested the proof of the duality Theorem 6.1.1. This new characterization of Sidon sets is proved by combining the duality theorem with a result of Rider [51] which connects Sidon sets and random Fourier series.

In Chapter VII we return to the classical results mentioned in the beginning of this chapter and show how they can be derived from our results. Moreover, we prove a general theorem on the rearrangements of the coefficients of an a.s. continuous random Fourier series. This implies in particular that if the series $\sum_{n=1}^{\infty} a_n \varepsilon_n e^{int}$ is a.s. continuous, then the series $\sum_{n=1}^{\infty} a_n^* \varepsilon_n e^{int}$ is also a.s. continuous where $\{a_n^*\}$ is the decreasing rearrangement of the sequence $\{|a_n|\}$. We consider random Fourier series that are bounded on the whole (non-compact) group and also show that the space $C_{a.s.}(G)$ is different if we take the process to be translation invariant on the right instead of on the left. Lastly we mention some generalizations of random Fourier series that can be studied by the methods given here and refer to some recent work along this line by ourselves and others.

We hope that these results will be of interest to probabilists who can view random Fourier series as interesting and useful examples of stochastic processes—intimately related to stationary Gaussian processes—and to harmonic analysts who may find these results and methods useful in their own work. Attempting to communicate to such different audiences has forced us to be repetitious but it is likely that different readers will concentrate on different parts of the book. For example a probabilist may wish to read only the first four chapters and parts of Chapter VII. If he is interested in harmonic analysis but is not familiar with the non-Abelian case then he can skip Chapter V and read Chapter VI assuming that the group is Abelian. On the other hand, if the reader is an analyst with strong interest in non-Abelian groups, we advise him to read the introduction and then to skip to Chapters V, VI and VII referring to the first three chapters when necessary.

The notation (a.b) refers to Section a, number b in the chapter in which it appears, whereas (a.b.c) refers to Chapter a, Section b, number c; this is only used when the material referred to is in another chapter.

We would like to thank our typist Ms. Vicki Davis.

CHAPTER II

PRELIMINARIES

1. *Covering number of compact metric spaces*

Suppose that we have a translation invariant metric on a locally compact Abelian group. Then there is a numerical equivalence between the number of open balls of radius ε (with respect to the metric) that is necessary to cover a certain compact subset of the group and the Haar measure of the trace on the compact subset, of an open ball of radius ε centered at the unit element of the group. This result, which will be given in Lemma 1.1, was suggested by Lemma 2.1 [23].

Let (S,ρ) be a compact metric or pseudo-metric space. We denote by $N_\rho(S,\varepsilon)$ the minimum number of open balls of radius ε in the metric or pseudo-metric ρ, with centers in S, that is necessary to cover S. Suppose that ρ is translation invariant on G and set $\delta(x) = \rho(0,x)$. (We employ here and in the following the notation introduced in Chapter I.) Define

$$m_\delta(\varepsilon) = \mu(x \in K \oplus K \,|\, \delta(x) < \varepsilon) \tag{1.1}$$

LEMMA 1.1. *In the above notation we have*

$$N_\rho(K \oplus K, \varepsilon) \le \frac{\mu_4}{m_\delta(\varepsilon/2)} \tag{1.2}$$

and

$$N_\rho(K, \varepsilon) \ge \frac{\mu_1}{m_\delta(\varepsilon)}\,. \tag{1.3}$$

Proof. Let $B(t,\varepsilon) = \{x \in G \,|\, \rho(t,x) \le \varepsilon\}$ and let $M_\rho(K \oplus K, \varepsilon)$ denote the maximum number of open balls of radius ε with respect to ρ, centered in $K \oplus K$ and disjoint in $\underset{4}{\oplus} K$. For all $t \in K \oplus K$ we have

$$\mu\{B(t, \varepsilon) \cap \underset{4}{\oplus} K\} \geq \mu\{B(0, \varepsilon) \cap K \oplus K\} \tag{1.4}$$

and

$$M_\rho(K \oplus K, \varepsilon/2) \geq N_\rho(K \oplus K, \varepsilon) . \tag{1.5}$$

Inequality (1.4) is elementary since for $t \in K \oplus K$, $t + \{B(0, \varepsilon) \cap K \oplus K\} \subset \{B(t, \varepsilon) \cap \underset{4}{\oplus} K\}$. Inequality (1.5) is a well-known fact which is proved as follows: Denote the centers of the $M_\rho(K \oplus K, \varepsilon/2)$ balls by $\{t_j;\ j = 1, \cdots, M_\rho(K \oplus K, \varepsilon/2)\}$. (These balls are not unique and if ρ is a pseudo-metric neither are the t_j but that doesn't affect the proof.) We have

$$K \oplus K \subset \bigcup_{j=1}^{M_\rho(K \oplus K, \varepsilon/2)} B(t_j, \varepsilon) . \tag{1.6}$$

To see this suppose (1.6) is false. Then there exists an $s \in K \oplus K$ but not contained in $\bigcup_{j=1}^{M_\rho(K \oplus K, \varepsilon/2)} B(t_j, \varepsilon)$. For such an s we have $\rho(s, t_j) > \varepsilon$ for all t_j. Let $u \in B(t_j, \varepsilon/2)$ then

$$\rho(s, u) \geq \rho(s, t_j) - \rho(t_j, u) > \varepsilon/2 .$$

Therefore $B(s, \varepsilon/2)$ is disjoint from $B(t_j, \varepsilon/2)$ for all j, $1 \leq j \leq M_\rho(K \oplus K, \varepsilon/2)$ and this contradicts the assumption that $M_\rho(K \oplus K, \varepsilon/2)$ is maximal. Thus we have established (1.6) and (1.5) follows immediately.

Using (1.4) and (1.5) we see that

$$\begin{aligned} \mu_4 &\geq \mu\Big(\bigcup_{j=1}^{M_\rho(K \oplus K, \varepsilon/2)} \{B(t_j, \varepsilon/2) \cap \underset{4}{\oplus} K\} \Big) \\ &\geq M_\rho(K \oplus K, \varepsilon/2)\, \mu(B(0, \varepsilon/2) \cap K \oplus K) \\ &\geq N_\rho(K \oplus K, \varepsilon)\, m_\delta(\varepsilon/2) \end{aligned}$$

which is (1.2).

To prove (1.3) we note that, analogous to (1.4), for all $t \in K$

$$\mu\{B(0,\varepsilon) \cap K \oplus K\} \geq \mu\{B(t,\varepsilon) \cap K\} . \tag{1.7}$$

Now suppose we have a minimal cover of K by balls of radius ε with respect to ρ with centers in K. There are $N_\rho(K,\varepsilon)$ balls in this cover; we denote their centers by $\{t_j,\ 1 \leq j \leq N_\rho(K,\varepsilon)\}$. Then, since $K \subset \bigcup_{j=1}^{N_\rho(K,\varepsilon)} B(t_j,\varepsilon)$, we have

$$\begin{aligned} \mu_1 &= \mu\Big\{\bigcup_{j=1}^{N_\rho(K,\varepsilon)} B(t_j,\varepsilon) \cap K\Big\} \\ &\leq N_\rho(K,\varepsilon)\,\mu\{B(0,\varepsilon) \cap K \oplus K\} \end{aligned}$$

where we use (1.7) at the last step. Thus we obtain (1.3).

REMARK 1.2. In the special case that G is compact and $K = G$ Lemma 1.1 is just

$$\frac{\mu(G)}{m_\delta(\varepsilon)} \leq N_\rho(G,\varepsilon) \leq \frac{\mu(G)}{m_\delta(\varepsilon/2)} \tag{1.8}$$

where $m_\delta(\varepsilon) = \mu(x \in G \mid \delta(x) < \varepsilon)$. This result is elementary and easy to prove directly.

In the next lemma we relate the covering number of $K \oplus K$ with the covering number of K. These observations are well known.

LEMMA 1.3. *Using the notation given above*

$$N_\rho(K, 2\varepsilon) \leq N_\rho(K \oplus K, \varepsilon) ; \tag{1.9}$$

$$N_\rho(K \oplus K, 2\varepsilon) \leq N_\rho^2(K,\varepsilon) . \tag{1.10}$$

Proof. Since K is a neighborhood of $0 \in G$, $K \subset K \oplus K$. The need for (1.9) is that the centers of the balls forming a cover of $K \oplus K$ may not be in K. Let $\{t_j, j = 1, \cdots, N_\rho(K\oplus K,\varepsilon)\}$ be the centers of balls of radius ε forming a cover of $K \oplus K$. For each j consider $\{B(t_j,\varepsilon) \cap K\}$. Clearly all of these sets are not empty. For each non-empty set choose an $s_j \in \{B(t_j, \varepsilon) \cap K\}$. The balls $\{B(s_j, 2\varepsilon)\}$ form a cover for K. Since, if not, there exists a $u \in K$ such that $\rho(s_j, u) \geq 2\varepsilon$ for all s_j. But then $\rho(t_j, u) \geq \rho(s_j, u) - \rho(s_j, t_j) > \varepsilon$, contradicting the fact that the $\{B(t_j, \varepsilon)\}$ cover K.

We now prove (1.10). Let $B(t_j, \varepsilon)$, $1 \leq j \leq N_\rho(K, \varepsilon)$ be a cover for K with centers $t_j \in K$. It follows that

$$\bigcup_{i,j=1}^{N_\rho(K,\varepsilon)} \{B(t_i, \varepsilon) \oplus B(t_j, \varepsilon)\}$$

is a cover for $K \oplus K$. Let $s \in B(t_i, \varepsilon) \oplus B(t_j, \varepsilon)$. Then $s = u_i + v_j$ for some $u_i \in B(t_i, \varepsilon)$ and $v_j \in B(t_j, \varepsilon)$. We have

$$\rho(s, t_i + t_j) \leq \rho(u_i, t_i) + \rho(v_j, t_j) < 2\varepsilon .$$

Therefore $\{B(t_i, \varepsilon) \oplus B(t_j, \varepsilon)\} \subset B(t_i + t_j, 2\varepsilon)$ and consequently

$$\bigcup_{i,j=1}^{N_p(K,\varepsilon)} B(t_i + t_j, 2\varepsilon)$$

is a cover for $K \oplus K$, with centers in $K \oplus K$, consisting of $N_\rho^2(K, \varepsilon)$ sets. This gives (1.10).

2. *A Jensen type inequality for the non-decreasing rearrangement of non-negative stochastic processes*

Let C be a compact subset of a locally compact Abelian group G and assume that G is not discrete. Let $g(x)$, $x \in C$ be a real valued non-negative measurable function on C. Define

$$\mu_g(\varepsilon) = \mu(x \in C \mid g(x) < \varepsilon)$$

where, as above, μ denotes the Haar measure of G. Set

$$\overline{g(u)} = \sup\{y \mid \mu_g(y) < u\} ;$$

$\overline{g}$ is called the non-decreasing rearrangement of g (with respect to C). Since $0 \le \mu_g \le \mu(C)$ the domain of $\overline{g}$ is the interval $[0, \mu(C)]$. We note that $\overline{g}$ viewed as a random variable on $[0, \mu(C)]$ has the same probability distribution with respect to normalized Lesbesgue measure on $[0, \mu(C)]$ that g has with respect to normalized Haar measure on C. In particular

$$\int_0^{\mu(C)} \overline{g(u)}\,du = \int_C g(x)\,\mu(dx) . \tag{2.1}$$

Lemma 2.1, which is a generalization of a well-known observation, provides an alternate definition of $\overline{g}$.

LEMMA 2.1. *For* $0 \le h \le \mu(C)$

$$\int_0^{h} \overline{g(u)}\,du = \inf_{\substack{E \subset C \\ \mu(E)=h}} \int_E g(x)\,\mu(dx) \tag{2.2}$$

i.e. the infimum is taken over all μ *measurable subsets* E *of* C *for which* $\mu(E) = h$. *Also, for* D *a non-negative constant we have*

$$\int_0^{h} \overline{Dg(u)}\,du = D \int_0^{h} \overline{g(u)}\,du . \tag{2.3}$$

Proof. Let E be a measurable subset of C, $\mu(E) = h$. Let $\tilde{\mu}(\varepsilon) = \mu\{x \in E \mid g(x) < \varepsilon\}$. Then $\tilde{\mu}(\varepsilon) \le \mu_g(\varepsilon)$ and

$$\int_0^{h} \overline{g(u)}\,du \le \int_0^{h} \sup\{y \mid \tilde{\mu}(y) < u\}\,du = \int_E g(x)\,\mu(dx)$$

where the last step is a statement of (2.1).

We complete the proof of (2.2) by exhibiting a set $F \subset C$ such that $\mu(F) = h$ and for which equality is attained in (2.2). This is quite simple when $\lambda\{u \in [0,h] | \overline{g(h)} = \overline{g(u)}\} = 0$, where λ denotes Lesbesgue measure. We consider this case first. Since g and $\overline{g}$ are equally distributed, as random variables, on their respective probability spaces, the probability distribution of $\overline{g(u)}$, for $u \in [0,h)$ with respect to normalized Lesbesgue measure is the same as the probability distribution of $g(x)$, for $x \in \{0 \leq g(x) < \overline{g(h)}\}$ with respect to normalized Haar measure. This implies that

$$(2.4) \qquad \int_0^h \overline{g(u)}\,du = \int_{\{0 \leq g(x) < \overline{g(h)}\}} g(x)\,\mu(dx)$$

where we also use the fact that $\lambda\{u \in [0,h] | \overline{g(h)} = \overline{g(u)}\} = 0$. The domain of integration of the integral on the right is the set F.

Now let $\lambda\{\overline{g(h)} = \overline{g(u)}\} = \mu\{g(x) = \overline{g(h)}\} = \delta \geq 0$. Let $h_0 = \sup_y \{\overline{g(y)} < \overline{g(h)}\}$. Then, as in (2.4),

$$\int_0^{h_0} \overline{g(u)}\,du = \int_{\{0 \leq g(x) < \overline{g(h)}\}} g(x)\,\mu(dx)\ .$$

Let $h - h_0 = \delta_1 \leq \delta$, so $\delta_1 \geq 0$. We can always decompose the set $\{x \in C | g(x) = \overline{g(h)}\}$ into two disjoint sets F_1 and F_2 with $\mu(F_1) = \delta_1$. Therefore we can write

$$\int_0^h \overline{g(u)}\,du = \int_{\{0 \leq g(x) < \overline{g(h)}\} \cup F_1} g(x)\,\mu(dx)\ .$$

The domain of integration of the integral on the right is the set F. This completes the proof of (2.2). The equality (2.3) follows immediately from (2.2).

The next lemma is also a generalization of a well-known observation. It displays the property of the non-decreasing rearrangement which lies behind Lemma 2.3. Its proof is immediate from Lemma 2.1.

LEMMA 2.2. *Let* $g(x)$ *and* $f(x)$ *be non-negative measurable functions on* C. *Then for* $0 \le h \le \mu(C)$

$$\int_0^h \overline{g+f}(x)\,\mu(dx) \ge \int_0^h \overline{g}(x)\,\mu(dx) + \int_0^h \overline{f}(x)\,\mu(dx)\,. \tag{2.5}$$

We now present the main result of this section. Let $(\Omega, \mathfrak{F}, P)$ be a probability space and let $\|\ \|_\Omega$ be a norm on the linear space of real valued functions on $(\Omega, \mathfrak{F}, P)$ with the property that if $|Y(\omega)| \ge |X(\omega)|$ a.s. (P), $\omega \in \Omega$, then $\|Y\|_\Omega \ge \|X\|_\Omega$. Let $g(x,\omega)$, $x \in C$ be a non-negative measurable function on $C \times \Omega$. The following lemma generalizes Lemma 1.1 [36].

LEMMA 2.3. *Let* $g(x,\omega)$ *and* $\|\ \|_\Omega$ *be as above and assume that* $\|g(x,\cdot)\|_\Omega < \infty$ *for* $x \in C$. *Let* $f(u) \ge 0$ *be a non-increasing function for* $0 \le u \le \mu(C)$ *and set* $\mu = \mu(C)$. *Then*

$$\Big\| \int_0^\mu \overline{g(u,\omega)}\, f(u)\,du \Big\|_\Omega \le \int_0^\mu \overline{\|g(u,\cdot)\|_\Omega}\, f(u)\,du\,. \tag{2.6}$$

In particular, when $f = I_{[0,h]}$, $0 \le h \le \mu$,

$$\Big\| \int_0^h \overline{g(u,\omega)}\,du \Big\|_\Omega \le \int_0^h \overline{\|g(u,\cdot)\|_\Omega}\,du\,. \tag{2.7}$$

Proof. We will first obtain (2.7). By Lemma 2.1, for each $\omega \in (\Omega, \mathfrak{F}, P)$,

$$\int_0^h \overline{g(u,\omega)}\,du \le \int_E g(x,\omega)\,\mu(dx)$$

for all $E \subset C$ with $\mu(E) = h$. Therefore

$$\| \int_0^h \overline{g(u,\omega)}\,du \|_\Omega \le \int_E \|g(x,\cdot)\|_\Omega\, \mu(dx)$$

and since this is true for all $E \subset C$, $\mu(E) = h$,

$$\| \int_0^h \overline{g(u,\omega)}\,du \|_\Omega \le \inf_{\substack{E \subset C \\ \mu(E)=h}} \int_E \|g(x,\cdot)\|_\Omega\, \mu(dx)\,. \tag{2.8}$$

By Lemma 2.1, the right side of (2.8) is equal to the right-hand side of (2.7).

There is nothing to prove in (2.6) unless the right side is finite. In this case the integral on the left in (2.6) is finite on a set $\Omega' \subset \Omega$, $P(\Omega') = 1$. This implies, since $\| \quad \|_\Omega$ is (by assumption) a lattice norm, that

$$\| \lim_{u \to 0} \int_0^u \overline{g(v,\omega)}\,dv f(u) \|_\Omega = 0\,. \tag{2.9}$$

The finiteness of the right side of (2.6) also implies that

$$\lim_{u \to 0} \int_0^u \overline{\|g(v,\cdot)\|_\Omega}\, dv f(u) = 0\,. \tag{2.10}$$

Integrating by parts and using (2.9) and (2.10) we have

$$\| \int_0^\mu \overline{g(u,\omega)} f(u)\,du \|_\Omega \le \| \int_0^\mu \overline{g(u,\omega)}\,du \|_\Omega f(\mu) \tag{2.11}$$

$$+ \| \int_0^\mu \int_0^u \overline{g(v,\omega)}\,dv \; d(-f(u)) \|_\Omega\,.$$

Using (2.7) and another integration by parts we have

$$(2.12)\qquad \left\| \int_0^{\mu}\int_0^{u} \overline{g(v,\omega)}\,dv\, d(-f(u))\right\|_{\Omega} \leq \int_0^{\mu}\int_0^{u} \overline{\|g(v,\cdot)\|_{\Omega}}\,dv\, d(-f(u))$$

$$= -\int_0^{\mu} \overline{\|g(v,\cdot)\|_{\Omega}}\,dv\, f(\mu) + \int_0^{\mu} \overline{\|g(u,\cdot)\|_{\Omega}}\, f(u)\,du\ .$$

Combining (2.11) and (2.12) and using (2.7) again we get (2.6).

The above results in this section follow from [36] and [25]. We will also record here an inequality of the nature of Lemma 2.3 due to Fernique (Proposition 1.4.2 [10]).

LEMMA 2.4. *Let* $X(t,\omega)$ *be a positive random variable on the product space of two probability spaces* $(\Omega,\mathfrak{F},P)\times(T,\mathfrak{T},\mu)$ *and* $a(\omega)$, $\omega\in\Omega$, *a positive random variable on* $(\Omega,\mathfrak{F},P)$. *Let* $\phi:[0,1]\to R^+$ *be convex and decreasing and let* E *be expectation with respect to* $(\Omega,\mathfrak{F},P)$. *Then*

$$E\left[\int_0^{a(\omega)} \phi\circ\mu\{t: X(t,\omega)<u\}\,du\right]$$

$$\leq \int_0^{Ea(\omega)} \phi\circ\mu\{t: EX(t,\omega)<u\}\,du\ .$$

3. *Continuity of Gaussian and sub-Gaussian processes*

Let $\{X(t), t\in U\}$, U an arbitrary index set, be a real valued stochastic process. The process is said to have sub-Gaussian increments if there exists a $\delta\geq 1$ such that for all $s,t\in U$ and $\lambda>0$

$$(3.0)\qquad E\{\exp(\lambda(X(s)-X(t)))\} \leq \exp\{\lambda^2\delta^2 E(X(t)-X(s))^2/2\}\ .$$

Gaussian processes satisfy this inequality with $\delta=1$. We will use a generalization of this definition. We say that a process has sub-Gaussian increments with respect to a metric (or pseudo-metric) $\rho(s,t)$, $s,t \in U$ if for all $s,t \in U$ and $\lambda > 0$

$$E\{\exp(\lambda(X(s)-X(t)))\} \leq \exp\{\lambda^2\rho^2(s,t)/2\}. \tag{3.1}$$

Obviously, (3.1) encompasses (3.0). It follows from (3.1) that $EX(t)$ is constant and

$$P\left[\frac{|X(s)-X(t)|}{\rho(s,t)} > u\right] \leq 2\exp(-u^2/2) \quad \text{for all} \quad u > 0. \tag{3.2}$$

(Cf. Lemma 5.2, Chapter II [25].) Conversely, any process $X(t)$ satisfying (3.2) and such that $EX(s) = EX(t)$ for all t, s in U, verifies (3.1) at least with ρ replaced by $K\rho$ for some numerical constant K.

As in Section 1 we let (S,ρ) denote a compact metric or pseudo-metric space and denote by $N_\rho(S,\varepsilon)$ the minimum number of balls of radius ε in the metric or pseudo-metric ρ, with centers in S, that cover S. The following theorem is an extension of Dudley's basic theorem [6] on sample path continuity of Gaussian processes; it is similar to a theorem of Fernique [9]; see also Theorem 4.5.5 [25].

THEOREM 3.1. *Let* $\tilde{S} = \{\tilde{X}(t), t \in T\}$, T *a compact topological space, be a stochastic process defined on* $(\Omega, \mathcal{F}, P)$ *and satisfying (3.1), where* $\rho(s,t)$ *is a continuous metric or pseudo-metric on* $T \times T$. *Define* $\hat{\rho} = \sup_{s,t\in T} \rho(s,t)$ *and assume that*

$$J(T,\rho) = J(\rho) = \int_0^{\hat{\rho}} (\log N_\rho(T,u))^{1/2} du < \infty. \tag{3.3}$$

Then there exists a version $S = \{X(t), t \in T\}$ *of the process with continuous sample paths satisfying the inequalities*

$$\{E[\sup_{\substack{\rho(s,t)\leq d\\ s,t\in T}} |X(s)-X(t)|^2]\}^{1/2} \tag{3.4}$$

$$\leq K\left[\int_0^d (\log N_\rho(T,u))^{1/2} du + \hat{\rho}\phi(d/4\hat{\rho})\right]$$

for all $0 \le d \le \hat{\rho}$ *and some absolute constant* $K \ge 1$, *where* $\phi(u) = u(\log\log 1/u)^{1/2}$ *and*

$$\{E[\sup_{t \in T} |X(t)|^2]\}^{1/2} \le K[\{E|X(t_0)|^2\}^{1/2} + \hat{\rho} + J(T,\rho)] \tag{3.5}$$

for any $t_0 \in T$. *(Note that (3.3) is unchanged if* $\hat{\rho}$ *is replaced by* ∞.)

Proof. Let $N_\rho(T,\varepsilon)$ denote the minimum number of open balls of radius ε in the metric or pseudo-metric ρ with centers in T that covers T. By (3.3) $N_\rho(T,\varepsilon) < \infty$ for all $\varepsilon > 0$. Let $\tilde{Y}(t) = \tilde{X}(t)/4\hat{\rho}$ and define $\tau(s,t) = \rho(s,t)/4\hat{\rho}$. By (3.2)

$$P[|\tilde{Y}(t) - \tilde{Y}(s)| > u] \le 2\exp\left[-1/2\left(\frac{u}{\tau(s,t)}\right)^2\right].$$

Note that $N_\tau(T,\varepsilon) = N_\rho(T, 4\hat{\rho}\varepsilon)$ is also finite for all $\varepsilon > 0$.

For all n there exist subsets $A_n \subset T$ such that $\text{card}(A_n) \le N(T,2^{-n})$ and given any $s \in T$ there exists a $t \in A_n$ such that $\tau(s,t) \le 2^{-n}$. We have

$$P(\{\sup_{\substack{s,t \in A_n \cup A_{n-1} \\ \tau(s,t) \le 2^{-n+2}}} |\tilde{Y}(s) - \tilde{Y}(t)| > b_n\}) \le 8N_\tau^2(T,2^{-n}) \cdot \exp[-b_n^2 2^{2n-5}] \equiv g(n). \tag{3.6}$$

Let Λ_n be the set in (3.6), then for $n_0 \ge 2$

$$P\{\bigcup_{n=n_0}^{\infty} \Lambda_n\} \le \sum_{n=n_0}^{\infty} g(n) \equiv G(n_0). \tag{3.7}$$

We will choose $\{b_n\}$ below such that $\Sigma b_n < \infty$ and $G(n_0) \downarrow 0$ as $n_0 \to \infty$.

For any $s \in T$ there exists a sequence $\{s_n\}$, $s_n \in A_n$ such that $\lim_{n\to\infty} \tau(s,s_n) = 0$. Furthermore given any $\varepsilon > 0$ there exists an n_0' such that $P\{\bigcup_{n=n_0'}^{\infty} \Lambda_n\} < \varepsilon$ and for $\omega \notin \bigcup_{n=n_0'}^{\infty} \Lambda_n$ and all $n_0 \ge n_0'$

$$\sup_{n,m\geq n_0} |\tilde{Y}(s_n,\omega)-\tilde{Y}(s_m,\omega)| \leq \sum_{n=n_0}^{\infty} b_n .$$

For these ω we define $Y(s,\omega)$ as the limit of the appropriate Cauchy sequence $\tilde{Y}(s_n,\omega)$.

Consider $s,t \in T$ such that $\tau(s,t) \leq 2^{-n_0}$, $n_0 \geq n_0'$. We will show that if $\omega \notin \bigcup_{n=n_0'}^{\infty} \Lambda_n$, then

(3.8) $$|Y(s,\omega)-Y(t,\omega)| \leq 3 \sum_{n=n_0}^{\infty} b_n .$$

This shows that the function $Y(s,\omega)$ just defined is uniformly continuous on T. Since this can be done for all $\varepsilon > 0$ we obtain a continuous separable version $\{Y(t), t \in T\}$ of $\{\tilde{Y}(t), t \in T\}$ and, as a trivial consequence, S of $\tilde{S}$.

To obtain (3.8) consider $s,t \in T$ such that $\tau(s,t) \leq 2^{-n_0}$. Also note that for all $n \geq n_0$ there exist s_n and $t_n \in A_n$ such that $\tau(s,s_n) \leq 2^{-n}$ and $\tau(t,t_n) \leq 2^{-n}$. Hence $\tau(s_{n_0},t_{n_0}) \leq 2^{-n_0+2}$, $\tau(s_n,s_{n+1}) \leq 2^{-n+1}$ and $\tau(t_n,t_{n+1}) \leq 2^{-n+1}$. Therefore, if $\omega \notin \bigcup_{n=n_0'}^{\infty} \Lambda_n$

(3.9) $$|Y(s,\omega)-Y(t,\omega)| \leq |\tilde{Y}(s_{n_0},\omega)-\tilde{Y}(t_{n_0},\omega)| + \sum_{n=n_0}^{\infty} |\tilde{Y}(t_n,\omega)-\tilde{Y}(t_{n+1},\omega)| + \sum_{n=n_0}^{\infty} |\tilde{Y}(s_n,\omega)-\tilde{Y}(s_{n+1},\omega)|$$

from which we get (3.8).

We also have, by (3.7) and (3.8)

(3.10) $$P\Big\{ \sup_{\tau(s,t)\leq 2^{-n_0}} |Y(s)-Y(t)| \geq 3 \sum_{n=n_0}^{\infty} b_n \Big\} \leq G(n_0) .$$

For $\lambda \geq 1$ we set

$$b_n^2 = \frac{\lambda^2}{2^{2n-8}} [\log N_\tau(T, 2^{-n-1}) + \log n]$$

then

$$(3.11) \quad \sum_{n=n_0}^{\infty} b_n \leq \lambda 2^5 \left[\sum_{n=n_0}^{\infty} \frac{(\log N_\tau(T, 2^{-n-1}))^{1/2}}{2^{n+1}} + \sum_{n=n_0}^{\infty} \frac{(\log n)^{1/2}}{2^{n+1}} \right]$$

$$\leq \lambda 2^5 \left[2 \int_0^{2^{-n_0-1}} (\log N_\tau(T, u))^{1/2} du + \frac{1}{2^{n_0-1}} (\log n_0)^{1/2} \right].$$

Substituting b_n^2 into $G(n_0)$ we get

$$(3.12) \quad G(n_0) = 8 \sum_{n=n_0}^{\infty} N_\tau^2(T, 2^{-n}) \exp[-8\lambda^2(\log N_\tau(T, 2^{-n-1}) + \log n)]$$

$$\leq 8 \exp[-6 \log N_\tau(T, 2^{-n_0-1})] \sum_{n=n_0}^{\infty} \exp[-8\lambda^2 \log n] .$$

The inequalities (3.11) and (3.12) prove our assertions about Σb_n and $G(n_0)$.

Let

$$Z = \sup_{\tau(s,t) \leq 2^{-n_0}} |Y(s) - Y(t)| \left[\frac{192}{\log 2} \int_0^{2^{-n_0-1}} (\log N_\tau(T, u))^{1/2} du + \frac{96}{2^{n_0-1}} (\log n_0)^{1/2} \right]^{-1} .$$

Then using (3.10), (3.11) and (3.12) we get

$$(3.13) \quad P[Z > \lambda] \leq 8 \exp[-6 \log N_\tau(T, 2^{-n_0-1})] \sum_{n=n_0}^{\infty} \exp[-8\lambda^2 \log n] .$$

This last inequality shows that $E[\exp \alpha Z^2] < \infty$ for some $\alpha > 0$ but for our purposes we need only note that

$$EZ^2 \leq 1 + \int_1^\infty 2\lambda P[Z > \lambda] d\lambda \leq 2 .$$

Thus

$$\{E[\sup_{\tau(s,t) \leq 2^{-n_0}} |Y(s) - Y(t)|^2]\}^{1/2}$$

$$\leq K' \left[\int_0^{2^{-n_0-1}} (\log N_\tau(T,u))^{1/2} du + \frac{(\log n_0)^{1/2}}{2^{n_0+1}} \right]$$

for some fixed constant K'. Extrapolating, we have for all $u \leq 1/4$

(3.14) $$\{E[\sup_{\tau(s,t) \leq u} |Y(s) - Y(t)|^2]\}^{1/2}$$

$$\leq K \left[\int_0^u (\log N_\tau(T,u))^{1/2} du + \phi(u) \right]$$

for a fixed constant K. Our main result (3.4) follows immediately by substituting $Y(t) = X(t)/4\hat{\rho}$ and making the change of variables in the integral using $N_\tau(T,\varepsilon) = N_\rho(T, 4\hat{\rho}\varepsilon)$. Note also that $\tau(s,t) \leq u$ implies $\rho(s,t) \leq 4\hat{\rho}u$ and the range of $4\hat{\rho}u$ is between 0 and $\hat{\rho}$. Inequality (3.5) follows immediately from (3.4) with $d = \hat{\rho}$. This completes the proof of Theorem 3.1.

Theorem 3.1 gives a sufficient condition for continuity of processes satisfying (3.1). A theorem of Fernique states that this condition is also necessary for stationary Gaussian processes. Fernique's theorem (Theorem 8.1.1 [9]) is proved for real valued processes on $\mathbf{R}^n$. We will mention here some minor modifications that are needed to apply it to the stationary Gaussian processes that are of interest to us. Let $\{\tilde{g}_\gamma\}_{\gamma\epsilon\Gamma}$ be a collection of independent normalized complex valued normal random variables. (The definition of $\tilde{g}$ follows (1.1.9).) If we consider the process

$$\tilde{H}(x) = \sum_{\gamma\epsilon A} a_\gamma \tilde{g}_\gamma \gamma(x) \qquad x \epsilon K , \tag{3.15}$$

then $\{\tilde{H}(x), x \epsilon K\}$ is a stationary complex valued Gaussian process. As will be explained in the next section, it is easy to see that the study of the process $\tilde{H}$ reduces to the study of

$$H(x) = \sum_{\gamma\epsilon A} a_\gamma g_\gamma \gamma(x) \qquad x \epsilon K , \tag{3.16}$$

where $\{g_\gamma\}_{\gamma\epsilon\Gamma}$ are i.i.d real normal Gaussian variables. We have then

$$\rho(x,y) = (E|H(x)-H(y)|^2)^{1/2} = \rho(x-y,0) = \sigma(x-y) ,$$

where σ is defined as in 1.1.12. Let $\hat{\sigma} = \sup\{\sigma(x) | x \epsilon K \oplus K\}$. To emphasize that we are dealing with translation invariant metrics, we will denote $J(K,\rho)$ also by $J(K,\sigma)$.

We will work with the following version of Fernique's theorem:

THEOREM 3.2. *A necessary condition for the process (3.16) to have a version with continuous sample paths is that* $J(K,\sigma) < \infty$.

Proof. Instead of considering $H(x)$ it is sufficient to prove the theorem for the real valued process

$$Y(x) = \sum_{\gamma\epsilon A} g_\gamma \operatorname{Re}(a_\gamma \gamma(x)) + \sum_{\gamma\epsilon A} g'_\gamma \operatorname{Im}(a_\gamma \gamma(x)), \quad x \epsilon K , \tag{3.17}$$

where $\{g'_\gamma | \gamma \epsilon A\}$ is an independent copy of $\{g_\gamma | \gamma \epsilon A\}$, since $E(H(x)-H(y))^2 = E(Y(x)-Y(y))^2 = \sigma^2(x-y)$ and the series (3.16) and (3.17) either both converge uniformly a.s. or neither does. Therefore by Theorem

4.3 below, the two series either both have versions with continuous sample paths or neither does.

The only point in the proof of Theorem 8.1.1 [9] that needs to be extended is Lemma 8.1.2. Let $H = \{x \in G | \sigma(x) = 0\}$ and form the quotient group $G' = G/H$. There exists a canonical mapping of G onto G'; let K' be the image of K under this mapping. Denote by σ' the metric on K' that corresponds to the pseudo-metric σ on K.

LEMMA 3.3. *There exists a* $\delta_0 > 0$ *and a compact symmetric neighborhood* S *of* 0, $S \subset K'$, *such that if* $s, t \in \underset{4}{\oplus} S$ *then* $\sigma'(s-t) \le \delta_0$ *implies* $s - t \in S$.

Proof. Let S be a compact symmetric neighborhood of $0 \in K'$ such that $\underset{8}{\oplus} S \subset K'$. Let $\beta = \min\{\sigma'(x), x \in \underset{8}{\oplus} S/S\}$. Since 0 is the unique zero of σ' on K' we have $\beta > 0$. Let $s, t \in \underset{4}{\oplus} S$ then $s - t \in \underset{8}{\oplus} S$. Set $\delta_0 = \beta/2$ then $\sigma'(s-t) \le \delta_0$ implies $s - t \in S$.

Consider S as given in Lemma 3.3 and let $T = \underset{4}{\oplus} S$. Following the notation of Theorem 8.1.1 [9] we define $B(S, \delta_0) = \bigcup_{s \in S} B(s, \delta_0)$ where $B(s, \delta)$ denotes an open ball of radius δ in K' with respect to the σ' metric. Let $s, t \in B(S, \delta_0) \cap T$, we show that for $\delta \le \delta_0$, $B(s, \delta) \cap T = A_1$ and $B(t, \delta) \cap T = A_2$ are translates of each other, i.e. if $u \in A_1$ then $u + t - s \in A_2$. To do this we need only show that $u + t - s \in T$. Since $t \in B(S, \delta_0)$ there exists a $t' \in S$ such that $\sigma(t-t') < \delta_0$. Set

$$u + t - s = t' + (t-t') + (u-s) .$$

Since $t, t' \in T = \underset{4}{\oplus} S$, by Lemma 3.3, $t - t' \in S$. Similarly $u - s \in S$ and since $t' \in S$ we have $u + t - s \in T$.

Consider the process

$$Y'(x) = \sum_{\gamma \in A} g_\gamma \operatorname{Re}(a_\gamma \gamma(x)) + \sum_{\gamma \in A} g'_\gamma \operatorname{Im}(a_\gamma \gamma(x)), \qquad x \in K'. \tag{3.18}$$

This is a real valued stationary Gaussian process with $(E|Y'(x)-Y'(y)|^2)^{1/2} = \sigma'(x-y)$ and an equivalent of Lemma 8.1.2 [9] holds for this process.

Assume that the series (3.18) has a version with continuous sample paths. Then it converges uniformly a.s. (see Theorem 4.3). By the Landau, Shepp, Fernique Theorem (Theorem 3.8 below) we have $E(\sup_{x \epsilon K'} Y'(x)) < \infty$. We refer to the second paragraph of 8.1.4 [9] with S and T as given above. This shows that there exists a $\delta' > 0$ such that

$$\int_0^{\delta'} (\log N_{\sigma'}(S,u))^{1/2} du < \infty$$

and since S is compact we also have $J(S,\sigma') < \infty$. Finally, since K' is compact, there exists a constant $C > 0$ such that $N_{\sigma'}(S,u) \geq CN_{\sigma'}(K',u)$. Therefore $J(K',\sigma') < \infty$. To obtain Theorem 3.2 for $Y(x)$, $x \epsilon K$ we note that the series (3.17) and (3.18) either both have versions with continuous sample paths or neither does. Furthermore

$$E(\sup_{x \epsilon K'} Y'(x)) = E(\sup_{x \epsilon K} Y(x))$$

and $N_{\sigma'}(K',u) = N_{\sigma}(K,u)$. Therefore we obtain Theorem 3.2.

When the group G is compact and $K = G$ we can obtain a two-sided bound for $E \sup_{x \epsilon G} |H(x)|$.

THEOREM 3.4. *Let* G *be compact,* $\mu(G) = 1$, *and take* $K = G$ *in (3.5), then*

(3.19)
$$C_0\left[\left(\sum_{\gamma \epsilon A} |a_\gamma|^2\right)^{1/2} + J(G,\sigma)\right] \leq E \sup_{x \epsilon G} |H(x)|$$
$$\leq C_1\left[\left(\sum_{\gamma \epsilon A} |a_\gamma|^2\right)^{1/2} + J(G,\sigma)\right]$$

where $0 < C_0, C_1 < \infty$ *are numerical constants independent of* $\{a_\gamma\}$ *and* $\{\gamma | \gamma \epsilon A\}$.

Proof. The upper bound is given by Theorem 3.1 since

$$\forall t_0 \epsilon G \quad (E|H(t_0)|^2)^{1/2} = \Big(\sum_{\gamma\epsilon A} |a_\gamma|^2\Big)^{1/2}$$

and

$$\rho(s,t) = \sigma(s-t) \le 2\Big(\sum_{\gamma\epsilon A} |a_\gamma|^2\Big)^{1/2} . \tag{3.20}$$

We now obtain the lower bound. We have trivially

$$E \sup_{x\epsilon G} Y(x) \le E \sup_{x\epsilon G} |Y(x)| . \tag{3.21}$$

Also both $E \sup_{x\epsilon G} |Y(x)|$ and $E \sup_{x\epsilon G} |H(x)|$ are greater than or equal to both $E \sup_{x\epsilon G} |\sum_{\gamma\epsilon A} g_\gamma \operatorname{Re}(a_\gamma \gamma(x))|$ and $E \sup_{x\epsilon G} |\sum_{\gamma\epsilon A} g_\gamma \operatorname{Im}(a_\gamma \gamma(x))|$. This remark is trivial with regard to $E \sup_{x\epsilon G} |H(x)|$; for $E \sup_{x\epsilon G} |Y(x)|$ it follows from Lemma 4.2 below. Therefore

$$\frac{1}{2} E \sup_{x\epsilon G} |H(x)| \le E \sup_{x\epsilon G} |Y(x)| \le 2E \sup_{x\epsilon G} |H(x)| . \tag{3.22}$$

Combining this with (3.21) we get

$$E \sup_{x\epsilon G} Y(x) \le 2E \sup_{x\epsilon G} |H(x)| . \tag{3.23}$$

The main ingredient in the proof is Fernique's Theorem 7.2.2 [9]. This theorem is not restricted to stationary Gaussian processes on a group. However, in this case it can be simplified and gives us the following inequality:

$$E \sup_{x\epsilon G} Y(x) \ge \delta/\sqrt{2\pi} \Big\{\sum_{n=1}^{\infty} 4^{-n}\{[\log_2 K_\delta(n)]\}^{1/2}\Big\} \tag{3.24}$$

for all $\delta > 0$. Here [ ] denotes "integral part" and $K_\delta(n)$ is the maximum number of disjoint open balls of radius $\delta/4^n$ centered in $B(0,\delta 4^{-n+1})$. The metric (or pseudo-metric) in this case is $\rho(s,t) = (E|Y(s)-Y(t)|^2)^{1/2} = \rho(s-t,0) = \sigma(s-t)$. (B is defined in the proof of Theorem 3.2.) For clarity, we now reproduce some of the computations in Fernique's paper. We define $m(\varepsilon) = m_\sigma(\varepsilon)$ as in (1.1.13) and note that

$$K_\delta(n)\, m(2\delta 4^{-n}) \geq m(\delta 4^{-n+1})$$

and also, by replacing δ by 2δ

$$K_{2\delta}(n)\, m(\delta 4^{-n+1}) \geq m(2\delta 4^{-n+1}) .$$

These inequalities are simple and follow the reasoning used in Lemma 1.1. They give us

$$\log_2 \frac{m(2\delta 4^{-n+1})}{m(2\delta 4^{-n})} \leq \log_2 K_\delta(n)\, K_{2\delta}(n) \tag{3.25}$$

Let $a_n = -\log_2 m(2\delta 4^{-n})$. Then using (3.25) in (3.24) we get

$$\sum_{n=1}^{\infty} 4^{-n}\{[a_n - a_{n-1}]\}^{1/2} \leq \frac{16\sqrt{\pi}}{\delta} E \sup_{x \in G} Y(x) + 1/3 . \tag{3.26}$$

Also

$$\begin{aligned}\sum_{n=1}^{\infty} 4^{-n}\{[a_n - a_{n-1}]\}^{1/2} &\geq \sum_{n=1}^{\infty} 4^{-n}(\sqrt{a_n - a_{n-1}} - 1) \\ &\geq \sum_{n=1}^{\infty} 4^{-n}(\sqrt{a_n} - \sqrt{a_{n-1}} - 1) \\ &\geq \frac{3}{4}\sum_{n=1}^{\infty} 4^{-n}\sqrt{a_n} - \sqrt{a_0}/4 - \frac{1}{3} .\end{aligned}$$

We set $\delta = \hat{\sigma}$ and obtain

$$\frac{1}{(\log 2)^{1/2}} \int_0^{\frac{\hat{\sigma}}{2}} \left(\log \frac{1}{m(\varepsilon)}\right)^{1/2} d\varepsilon \le 4\hat{\sigma}(\log 2) \sum_{n=1}^{\infty} 4^{-n}\sqrt{a_n}$$

and $a_0 = 0$. Thus

$$(3.27)\quad (1/24\sqrt{\pi})\hat{\sigma} + E \sup_{x \in G} Y(x) \ge \frac{3}{(16)^2(\log 2)^{3/2}\sqrt{\pi}} \int_0^{\hat{\sigma}/2} \left(\log \frac{1}{m(\varepsilon)}\right)^{1/2} d\varepsilon .$$

We note a homogeneity property of $\int_0^{\hat{\sigma}/2} \left(\log \frac{1}{m(\varepsilon)}\right)^{1/2} d\varepsilon$. To be explicit let us denote

$$\sigma_Y(u) = (E|Y(u) - Y(0)|^2)^{1/2}$$

and

$$m_{\sigma_Y}(\varepsilon) = \mu\{x \in G | \sigma_Y(x) < \varepsilon\}$$

and the above integral by

$$(3.28)\qquad \int_0^{\hat{\sigma}_Y/2} \left(\log \frac{1}{m_{\sigma_Y}(\varepsilon)}\right)^{1/2} d\varepsilon$$

where $\hat{\sigma}_Y = \sup_{u \in G} \sigma_Y(u)$. Now consider the process $g = Y/\|Y\|$ where $\|Y\| = E \sup_{x \in G} |Y(x)|$. It is easy to check that

$$\sigma_g = \sigma_Y/\|Y\| \quad \text{and} \quad m_{\sigma_g}(\varepsilon) = m_{\sigma_Y}(\|Y\|\varepsilon) .$$

Therefore the integral in (3.28) is equal to

$$(3.29)\qquad \|Y\| \int_0^{\hat{\sigma}_g/2} \left(\log \frac{1}{m_{\sigma_g}(\varepsilon)}\right)^{1/2} d\varepsilon .$$

We now return to (3.27). Suppose that $\|Y\| = 1$ then

$$\hat{\sigma} \le 2\Big(\sum_{\gamma \epsilon A} |a_\gamma|^2\Big)^{1/2} \le \sqrt{2\pi}\, E|Y(0)| \le \sqrt{2\pi}\,.$$

Therefore for processes $Y(x)$ with $E \sup_{x \epsilon G} |Y(x)| = 1$ we have

$$\int_0^{\hat{\sigma}/2} \Big(\log \frac{1}{m(\varepsilon)}\Big)^{1/2} d\varepsilon \le D$$

where $D \le \dfrac{(16)^2 (\log 2)^{3/2} \sqrt{\pi}}{3}\,(1 + \sqrt{2}\,\pi)$. By (3.29) we have, for general Y,

$$E \sup_{x \epsilon G} Y(x) \ge \frac{1}{D} \int_0^{\hat{\sigma}/2} \Big(\log \frac{1}{m(\varepsilon)}\Big)^{1/2} d\varepsilon$$

and by Remark 1.2 and (3.23)

$$E \sup_{x \epsilon G} |H(x)| \ge \frac{1}{4D}\, J(G, \sigma)$$

Also, since $E \sup_{x \epsilon G} |H(x)| \ge \sqrt{2/\pi}\Big(\sum_{\gamma \epsilon A} |a_\gamma|^2\Big)^{1/2}$ we obtain the left side of (3.19).

REMARK 3.5. We claim that for G compact

$$\Big(\sum_{\substack{\gamma \epsilon A \\ \gamma \ne 0}} |a_\gamma|^2\Big)^{1/2} \le \nu\, J(G, \sigma) \quad \text{for some constant } \nu\,.$$

Indeed, since $N(\varepsilon) \ge \dfrac{1}{m(\varepsilon)}$, we have

$$J(G, \sigma) \ge \int_0^\infty \Big(\log \frac{1}{m(\varepsilon)}\Big)^{1/2} d\varepsilon = \int_0^1 \Big(\log \frac{1}{t}\Big)^{1/2} d\bar{\sigma}(t)$$

$$\ge \Big(\log \frac{1}{\varepsilon}\Big)^{1/2} \int_0^\varepsilon d\bar{\sigma}(t) = \Big(\log \frac{1}{\varepsilon}\Big)^{1/2} \bar{\sigma}(\varepsilon)\,.$$

Hence $\int_0^1 \overline{\sigma}(\varepsilon)\,d\varepsilon \le \nu J(G,\sigma)$ with $\nu = \int_0^1 \left(\log \frac{1}{\varepsilon}\right)^{-1/2} d\varepsilon$. Now, since σ and $\overline{\sigma}$ have the same distribution, we have

$$\Big(\sum_{\substack{\gamma\epsilon A\\ \gamma\neq 0}} |a_\gamma|^2\Big)^{1/2} \le \int \sigma(t)\,dm(t) = \int_0^1 \overline{\sigma}(\varepsilon)\,d\varepsilon \le \nu J(G,\sigma)\ ;$$

and this proves the above claim. This inequality, together with (3.19), shows that if we assume that $0 \notin A$, then the functional

$$\{a_\gamma\}_{\gamma\epsilon A} \to J(G,\sigma)$$

is equivalent to a norm on the set of families $\{a_\gamma\}_{\gamma\epsilon A}$ for which it is finite. Curiously, we know of no direct proof of this fact.

We now show that the two integrals $I(\sigma)$ defined in (1.1.14) and $J(K,\sigma)$ defined in (3.3) (see also the statement preceding the proof of Theorem 3.2) are equivalent.

LEMMA 3.6. *Assume that* $J(\sigma) < \infty$, *then*

$$-C_1\hat{\sigma} + \frac{1}{2} I(\sigma) \le J(\sigma) \le C_2\hat{\sigma} + 2I(\sigma)\ , \tag{3.30}$$

where $0 < C_1, C_2 < \infty$.

Proof. We first note that by the change of variables $u = \overline{\sigma(s)}$ and the fact that $\overline{\sigma(\mu_2)} = \hat{\sigma}$

$$\int_0^{\hat{\sigma}} \left(\log \frac{4\mu_4}{m(u)}\right)^{1/2} du = \int_0^{\mu_2} \left(\log \frac{4\mu_4}{s}\right)^{1/2} d\overline{\sigma(s)} \tag{3.31}$$

$$= \left(\log \frac{4\mu_4}{\mu_2}\right)^{1/2} \hat{\sigma} + \frac{1}{2} I(\sigma)\ .$$

By (1.9) and Lemma 1.1 we have

$$(3.32) \qquad J(K,\sigma) \le 2J(K \oplus K, \sigma) \le 4 \int_0^{\hat{\sigma}} \left(\log \frac{4\mu_4}{m(u)}\right)^{1/2} du \,.$$

Combining (3.32) and (3.31) we get the right side of (3.30). For the left side we have by (1.3)

$$(3.33) \qquad J(\sigma) \ge \int_0^{\hat{\sigma}} \left(\log \frac{\mu_1}{m_\sigma(\varepsilon)}\right)^{1/2} d\varepsilon$$

$$\ge \int_0^{\hat{\sigma}} \left(\log \frac{4\mu_4}{m_\sigma(\varepsilon)}\right)^{1/2} d\varepsilon - \hat{\sigma}\left(\log \frac{4\mu_4}{\mu_1}\right)^{1/2}$$

$$\ge \frac{1}{2} I(\sigma) - \hat{\sigma}\left(\log \frac{\mu_2}{\mu_1}\right)^{1/2}$$

where at the last step we use (3.31).

We mention two more results from the theory of Gaussian processes. Not in their greatest generality but in the form that we need them. The first is Belyaev's dichotomy.

THEOREM 3.7. *Consider the Gaussian process* $H(x)$ *defined in (3.16). Either* $H(x)$ *has a version with continuous sample paths a.s. or else, for all open sets* $U \subset K$,

$$\sup_{x \in U} |H(x)| = \infty \text{ a.s.}$$

This theorem was proved by Belyaev [2] for stationary Gaussian processes on the real line. In this generality it follows from a result of Ito and Nisio. (See Theorem 4.7 and Theorem 4.9 in Chapter 3 of [25] and recall that K is metrizable. Use these theorems to obtain the result for

Re H(x) and Im H(x). (see (3.16)) both of which are stationary. The result then follows for H(x).)

The second is the Landau, Shepp, Fernique Theorem (Corollary 4.7, Chapter 2 [25]).

THEOREM 3.8. *Let* $Y(x), x \in K$ *be a real valued, mean zero, separable Gaussian process with continuous sample paths a.s., then there exists an* $\varepsilon > 0$ *such that*

$$E[\exp(\varepsilon \sup_{x \in K} |Y(x)|^2)] < \infty . \tag{3.34}$$

A different approach to obtain sufficient conditions for the continuity of Gaussian processes was introduced by Garsia, cf. [14]. C. Preston developed the method further in [50]. For the applications to Harmonic Analysis given in Chapter VI, it will be convenient to use the following result of Preston [50]:

THEOREM 3.9. *Let* $\psi : R_+ \to R_+$ *be a convex (increasing and continuous) function with* $\psi(0) = 0$. *Let* (X, d) *be a compact metric space and let* μ *be a probability measure on* X. *We set for all* $\varepsilon > 0$

$$m(\varepsilon) = \inf_{x \in X} \mu(\{y \in X | d(x,y) < \varepsilon\})$$

and $R = \sup\{d(x,y) | x, y \in X\}$. *Suppose that* $I = \int_0^R \psi^{-1}\left[\frac{1}{m(\varepsilon/2)^2}\right] d\varepsilon < \infty$. *Then, every* f *in* $L_2(X,\mu)$ *which satisfies*

$$\int \psi\left(\frac{|f(s)-f(t)|}{a d(s,t)}\right) \mu(ds)\, \mu(dt) \le 1$$

for some $a > 0$, *must be equal* μ *–a.e. to a continuous function which we again denote by* f *which verifies for all* $x, y \in X$

(3.35) $$|f(x)-f(y)| \le 20\alpha \int_0^{d(x,y)} \psi^{-1}\left[\frac{1}{m(\varepsilon/2)^2}\right] d\varepsilon .$$

As a consequence, we have

(3.36) $$\|f\|_{C(X)} \le \|f\|_{L_2(d\mu)} + 20\alpha I .$$

The preceding theorem reproduces Theorem 3 in [50] except for the last line. Note that (3.35) obviously implies

$$\|f\|_{C(X)} \le \inf\{|f(x)| \mid x \in X\} + 20\alpha I$$
$$\le \left(\int |f(x)|^2 \, d\mu(x)\right)^{1/2} + 20\alpha I .$$

Therefore (3.35) trivially implies (3.36).

4. *Sums of Banach space valued random variables*

We record here a number of well-known results concerning sums of random variables with values in a Banach space. In many of these results, the random variables are independent, but it will be useful to consider more generally sign-invariant sequences of random variables. A sequence $\{X_k\}_{k \in N}$ of Banach space valued random variables is called sign-invariant if for any choice of signs $\bar{\varepsilon}_k = \pm 1$ the sequence $\{\bar{\varepsilon}_k X_k\}$ has the same distribution as the original sequence $\{X_k\}$. Equivalently, let $\{\varepsilon_k\}_{k \in N}$ be a Rademacher sequence (we will also refer to it as a Bernoulli sequence) independent of $\{X_k\}$. The sequence $\{X_k\}$ is sign-invariant if and only if it has the same probability distribution as the sequence $\{\varepsilon_k X_k\}$. Thus, for example, if the variables $\{X_k\}$ are independent and symmetric, then $\{X_k\}$ is sign-invariant; but, of course, the converse is not true (see, e.g. Remark 3.1.3).

We define

(4.1) $$S_n = \sum_{k=0}^{n} X_k \quad \text{and} \quad M = \sup_n \|S_n\| ,$$

where of course $\| \ \|$ indicates the Banach space norm.

LEMMA 4.1. *Let* $\{X_k\}$ *be a sign-invariant sequence of random variables with values in a Banach space* B, *defined on a probability space* $(\Omega, \mathcal{F}, P)$.

(i) *Assume that the series* $S = \sum_{k=1}^{\infty} X_k$ *is convergent in norm a.s. Then for all* $c > 0$

$$P(\{\|S\| > c\}) \geq \frac{1}{2} P(\{M > c\}) \geq \frac{1}{2} P\left(\left\{\sup_k \|X_k\| > \frac{c}{2}\right\}\right). \tag{4.2}$$

$$\text{For all } p > 0, \quad EM^p \leq 2E\|S\|^p. \tag{4.2$'$}$$

(ii) *If, for some* $p > 0$, *the series* $S = \sum_{k=1}^{\infty} X_k$ *is convergent in* $L_p(dP; B)$ *then* S_n *converges a.s. to* S *and moreover*

$$E \sup_{m>n} \|S_m - S_n\|^p \to 0 \quad \text{when} \quad n \to \infty. \tag{4.2$''$}$$

The first inequality in (4.2) is Lévy's inequality; the proof in [26], p. 12 is for $\{X_k\}$ independent and symmetric but it works in this case also. The second inequality follows because $X_k = S_k - S_{k-1}$ implies $\sup \|X_k\| \leq 2M$.

The inequality (4.2′) follows immediately from (4.2) and from the well-known equality

$$EM^p = \int_0^{\infty} P(\{M > c\}) pc^{p-1} dc.$$

Finally, we have by (4.2′)

$$E \sup_{m>n} \|S_m - S_n\|^p \leq 2E\|S - S_n\|^p.$$

Therefore, if $E\|S - S_n\|^p \to 0$ then (4.2″) holds.

This clearly shows also that $S_n \to S$ a.s. when $n \to \infty$. As the reader may have noticed, for $p \geq 1$ the fact that $S_n \to S$ a.s. also follows from the martingale convergence theorem (cf. [43], Prop. V.2.6).

The next lemma relates a random variable to its symmetrization.

LEMMA 4.2. *Let* X *be a Banach space valued random variable with* $E\|X\| < \infty$ *and* $EX = 0$. *Let* X' *be a copy of* X *defined on an independent probability space* $(\Omega', \mathcal{F}', P')$. *Then, if* $p \geq 1$

$$(4.3) \qquad (E\|X\|^p)^{1/p} \leq (E\|X-X'\|^p)^{1/p} \leq 2(E\|X\|^p)^{1/p} .$$

Proof. For the left side of (4.3) we write

$$E\|X-X'\|^p = E_\omega E_{\omega'}\|X(\omega)-X(\omega')\|^p$$
$$\geq E_\omega\|X(\omega)-E_{\omega'}X'(\omega')\|^p = E\|X\|^p ,$$

where $\omega \in \Omega$, $\omega' \in \Omega'$ and $E_\omega, E_{\omega'}$ represent expectation with respect to the relevant probability space. The right side is trivial.

The next theorem is a well-known result due to Ito and Nisio [21].

THEOREM 4.3. *Let* S *be a compact metric (or pseudo-metric) space and denote by* $C(S)$ *the Banach space of continuous functions on* S *with the standard sup-norm. Let* $\{X_k\}$ *be a sequence of independent symmetric* $C(S)$*-valued random variables. Then the following are equivalent*:

(i) $\sum_{k=0}^{\infty} X_k(t)$ *converges for each* $t \in S$ *and the resulting process has a version with continuous sample paths, and*

(ii) $\sum_{k=0}^{\infty} X_k$ *converges uniformly a.s. (i.e. converges in* $C(S)$ *a.s.).*

For a proof, see e.g. [25], Theorem 3.4, Chapter II or [18], Theorem 6.2.

REMARK 4.4. The preceding statement is still valid assuming merely that $\{X_k\}$ is a sign-invariant sequence. This claim can be justified as follows: Let $\{\varepsilon_k\}$ be a Bernoulli sequence on a probability space $(\Omega_2, \mathcal{F}_2, P_2)$ and assume that $\{X_k\}$ is defined on a probability space $(\Omega_1, \mathcal{F}_1, P_1)$. Since

$\{X_k\}$ is sign invariant, $\{X_k\}$ and $\{\varepsilon_k X_k\}$ have the same probability distribution. Now, assume that $\sum \varepsilon_k X_k$ has a version with continuous sample paths. Then (using e.g. Lemma 6.1 in [18]) we see that this implies that for almost all $\omega_1 \in \Omega_1$ the process $\sum_1^\infty \varepsilon_k(\omega_2) X_k(\omega_1)$ has a version with continuous sample paths. Hence, by Theorem 4.3 above, for almost all $\omega_1 \in \Omega_1$, $\sum \varepsilon_k(\omega_2) X_k(\omega_1)$ converges uniformly a.s. with respect to $(\Omega_2, \mathcal{F}_2, P_2)$. It follows by Fubini's theorem that $\sum \varepsilon_k X_k$ converges uniformly a.s. with respect to $P_1 \times P_2$ and this justifies the above claim.

The next theorems of Kahane are fundamental.

THEOREM 4.5 (Theorem 4 in [26], p. 17). *Let* $\{u_k\}$ *be a sequence of elements of a Banach space* B, $\{\varepsilon_k\}$ *a Rademacher (or equivalently Bernoulli) sequence and assume that*

$$S = \sum_{k=0}^{\infty} \varepsilon_k u_k \quad \textit{converges a.s.}$$

(resp. $M = \sup_n \|\sum_1^n \varepsilon_k u_k\| < \infty$ *a.s.). Then, for all* $p > 0$, $E\|S\|^p < \infty$ *(resp.* $EM^p < \infty$ *). Moreover, for any fixed* $p > 0$ *the series* $\sum_1^\infty \varepsilon_k u_k$ *is convergent in* $L_p(B)$ *if and only if it is a.s. convergent.*

The reader may have noticed that the last assertion follows from the second part of Lemma 4.1.

COROLLARY 4.6. *With the above notation, we have for* $0 < p < q < \infty$,

$$\left(E\|\sum_{k=1}^{\infty} \varepsilon_k u_k\|^q\right)^{1/q} \le K_{pq}\left(E\|\sum_1^{\infty} \varepsilon_k u_k\|^p\right)^{1/p} \tag{4.4}$$

where K_{pq} *is a constant depending on* p *and* q *only.*

This corollary is an immediate consequence of Theorem 4.5 and the closed graph theorem, so that the inequalities (4.4) are usually referred to as Kahane's inequalities. For a direct proof of (4.4), see [46].

Theorem 4.5 has an analogue for Gaussian processes which follows from the Fernique-Landau-Shepp theorem (Theorem 3.8).

THEOREM 4.7. *Let* $\{u_k\}$ *be as above and let* $\{g_k\}$ *be a sequence of independent* $N(0,1)$ *random variables. If* $S = \sum_1^\infty g_k u_k$ *converges a.s. (resp. if* $M = \sup \|\sum_1^n g_k u_k\| < \infty$ *a.s.) then, necessarily,* $E\|S\|^p < \infty$ *(resp.* $EM^p < \infty$ *) for each* $p < \infty$. *Moreover, for each fixed* $p > 0$, *the series* $\sum_{k=1}^\infty g_k u_k$ *converges in* $L_p(B)$ *if and only if it converges a.s.*

The last assertion follows from the first one as indicated in the second part of Lemma 4.1.

Of course, one can also derive an analogue of (4.4) in the Gaussian case. In fact, it is even possible to deduce directly the corresponding inequality from (4.4) using the (finite dimensional) central limit theorem. We do this in the next corollary.

COROLLARY 4.8. *With the notation of Theorem 4.7, for* $0 < p < q < \infty$ *we have that for all* N

$$(4.5)\qquad \left(E\|\sum_1^N g_k u_k\|^q\right)^{1/q} \le K_{pq}\left(E\|\sum_1^N g_k u_k\|^p\right)^{1/p}.$$

(Note that this inequality also holds for $N = \infty$, *in the limit.)*

Proof. Let $m \ge 1$ be an arbitrary integer; a suitable application of (4.4) yields

$$\left(E\|\sum_{k=1}^N \sum_{j=1}^m 1/\sqrt{m}\, \varepsilon_{km+j} u_k\|^q\right)^{1/q} \le K_{pq}\left(E\|\sum_{k=1}^N \sum_{j=1}^m 1/\sqrt{m}\, \varepsilon_{km+j} u_k\|^p\right)^{1/p}.$$

Therefore, if N is fixed and if $m \to \infty$, the central limit theorem immediately gives (4.5).

In his monograph [26], Kahane made extensive use of what he named the "contraction principle." This principle has been studied in greater generality in the papers [18], [24], and [19]. In this paper, we will need only a very simple form of the contraction principle, as follows:

THEOREM 4.9. *Let* $\{u_k\}$ *be a sequence of elements in a Banach space* B. *Let* $\{\varepsilon_k\}$ *be a Rademacher sequence and let* $\{\xi_k\}$ *be a sign-invariant sequence of complex valued integrable random variables. Then for any integer* N *and for any* p *with* $1 \le p < \infty$,

$$(4.6) \qquad \left(E\left\|\sum_1^N \xi_k u_k\right\|^p\right)^{1/p} \le 2 \sup_{k\le N} \|\xi_k\|_\infty \left(E\left\|\sum_1^N \varepsilon_k u_k\right\|^p\right)^{1/p}$$

where $\|\xi_k\|_\infty$ *means* ess sup $|\xi_k(\cdot)|$, *and*

$$(4.7) \qquad (\inf_{k\le N} E|\xi_k|)\left(E\left\|\sum_1^N \varepsilon_k u_k\right\|^p\right)^{1/p} \le 2\left(E\left\|\sum_1^N \xi_k u_k\right\|^p\right)^{1/p}.$$

Proof. We may as well assume that $\{\xi_k\}$ is defined on a probability space $(\Omega_1, \mathcal{F}_1, P_1)$ and that $\{\varepsilon_k\}$ is defined on $(\Omega_2, \mathcal{F}_2, P_2)$. We denote E_1 and E_2 the corresponding expectations. We first claim that if $\{a_j\}_{j\le N}$ are real numbers with $\sup_{j\le N} |a_j| \le 1$, then

$$(4.8) \qquad \left(E\left\|\sum_1^N a_k\varepsilon_k u_k\right\|^p\right)^{1/p} \le \left(E\left\|\sum_1^N \varepsilon_k u_k\right\|^p\right)^{1/p}.$$

This follows by a simple convexity argument. Consider $\tilde{a} = \{a_k\}_{k=1,\cdots,N}$ as an element of $[-1,1]^N$. Then $\tilde{a} = \sum_j b_j \tilde{e}_j$ where $\tilde{e}_j$ is an extremal point of $[-1,1]^N$ and $b_j > 0$, $\sum b_j = 1$. Denote the function $f: [-1,1]^N \to R$ by $f(\tilde{a}) = \left(E\left\|\sum_{k=1}^N a_k\varepsilon_k u_k\right\|^p\right)^{1/p}$. Then $f(\tilde{a}) \le$

$\sum_j b_j f(\tilde{e}_j)$ and since $\sum b_j = 1$ we see that $f(\tilde{a}) \le f(\tilde{e}_j)$ for some j. We now have (4.8) since $f(\tilde{e}_j)$ is equal to the right side of (4.8). Therefore for a_j complex numbers we have

$$(4.9) \qquad \left(E\left\|\sum_1^N a_k u_k \varepsilon_k\right\|^p\right)^{1/p} \le 2\left(E\left\|\sum_1^N \varepsilon_k u_k\right\|^p\right)^{1/p}.$$

We can now prove (4.6); we set $\alpha = \sup_{k \le N} \|\xi_k\|_\infty$, then by (4.9) we have for any fixed $\omega_1 \in \Omega_1$

$$E_2\left\|\sum_1^N \varepsilon_k \xi_k(\omega_1) u_k\right\|^p \le 2^p \alpha^p E_2\left\|\sum_1^N \varepsilon_k u_k\right\|^p ;$$

integrating with respect to P_1, we obtain (4.6).

We now turn to (4.7); by (4.9) applied with $a_k = |\xi_k(\omega_1)|(\xi_k(\omega_1))^{-1}$ with ω_1 fixed and u_k replaced by $\xi_k(\omega_1) u_k$ with ω_1 fixed and $\xi_k(\omega_1) \ne 0$ we get

$$2^p E_2\left\|\sum_1^N \varepsilon_k \xi_k(\omega_1) u_k\right\|^p \ge E_2\left\|\sum_1^N \varepsilon_k |\xi_k(\omega_1)| u_k\right\|^p$$

and integrating

$$2^p E_1 E_2\left\|\sum_1^N \varepsilon_k \xi_k(\omega_1) u_k\right\|^p \ge E_1 E_2\left\|\sum_1^N \varepsilon_k |\xi_k(\omega_1)| u_k\right\|^p .$$

By convexity this is

$$(4.10) \qquad \ge E_2\left\|\sum_1^N \varepsilon_k E_1 |\xi_k| u_k\right\|^p .$$

Finally, we apply (4.8) with $a_k = (\inf_{k \le N} E_1|\xi_k|)(E_1|\xi_k|)^{-1}$ and with u_k replaced by $E_1|\xi_k| u_k$ to see that (4.10)

$$\geq \left(\inf_{k \leq N} E_1 |\xi_k|\right)^p E_2 \left\| \sum_1^N \varepsilon_k u_k \right\|^p .$$

Inequality (4.7) now follows because

$$E \left\| \sum_1^N \xi_k u_k \right\|^p = E_1 E_2 \left\| \sum_1^N \varepsilon_k \xi_k u_k \right\|^p .$$

REMARK 4.10. Obviously, when the variables $\{\xi_n\}$ are real valued, the preceding proof gives (4.6) and (4.7) without the factor 2.

REMARK 4.11. In the case when $\{\xi_n\}$ is a Steinhaus sequence, Theorem 4.9 shows that the Steinhaus series behave in an essentially equivalent way as the Rademacher series. For instance, it is easy to deduce from Proposition 4.9 an analogue of (4.4) for Steinhaus series.

LEMMA 4.12. *Let* $\{g_n\}$ *and* $\{\tilde{g}_n\}$ *be sequences of real and complex valued Gaussian random variables as defined in Chapter I. Let* $\{u_k\}$ *be as above. Then, for* $1 \leq p < \infty$, *for all* N

$$(4.11) \qquad (1/\sqrt{2})^p E \left\| \sum_1^N g_k u_k \right\|^p \leq E \left\| \sum_1^N \tilde{g}_k u_k \right\|^p \leq (\sqrt{2})^p E \left\| \sum_1^N g_k u_k \right\|^p$$

Proof. This is entirely elementary. Since $\{\mathrm{Re}\, \tilde{g}_k\}$ and $\{\mathrm{Im}\, \tilde{g}_k\}$ are independent and equally distributed and both have the same distribution as $\{1/\sqrt{2}\, g_k\}$, the left side of (4.11) follows from (4.3). The right side follows from the triangle inequality.

REMARK 4.13. For any complex valued mean zero Gaussian process $\{\tilde{X}(t), t \in T\}$, T a compact metric space, there is an associated real valued Gaussian process $\{X(t), t \in T\}$ satisfying, for all $s, t \in T$

$$(4.12) \qquad E|\tilde{X}(t) - \tilde{X}(s)|^2 = E|X(t) - X(s)|^2$$

and

(4.13) $$E|\tilde{X}(t)|^2 = E|X(t)|^2$$

such that $\tilde{X} = \{X(t), t \in T\}$ has a version with continuous sample paths if and only if $X = \{X(t), t \in T\}$ does. (Note that in this vein, (4.11) along with Theorem 4.7, shows the mutual convergence a.s. of $\sum_1^\infty g_k u_k$ and $\sum_1^\infty \tilde{g}_k u_k$.) To see this consider $\operatorname{Re} \tilde{X}(t)$ and $\operatorname{Im} \tilde{X}(t)$. Clearly $\tilde{X}$ has a version with continuous sample paths if and only if $\operatorname{Re} \tilde{X}$ and $\operatorname{Im} \tilde{X}$ have versions with continuous sample paths. We define X by

(4.14) $$X(t) = \operatorname{Re} \tilde{X}(t) + \operatorname{Im} \tilde{X}'(t), \quad t = T,$$

where $\tilde{X}' = \{\tilde{X}'(t), t \in T\}$ is an independent copy of $\tilde{X}$. It is easy to see that the above assertions are true.

REMARK 4.14. The above results will be used in Chapter III in the following way: We will consider a locally compact Abelian group G with dual Γ and apply Theorem 4.5 and what follows to the Banach space $C(K)$, where K is a compact metrizable subset of G. The elements u_k of B will be of the form $u_k = a_k \gamma_k$ with $\{a_k\}$ complex numbers and with $\{\gamma_k\}$ a sequence of elements of Γ considered as functions on K.

The next result uses the notation of the preceding section.

COROLLARY 4.15. *Let H be an a.s. continuous Gaussian process of the form (3.16). Then there is a constant β_K depending only on K and G such that*

(4.15) $$J(K, \sigma) \le \beta_K (E\|H\|^2)^{1/2}$$

where $\|\ \|$ is the norm in $C(K)$.

Proof. For $\tilde{H}$ as given in (3.15) we have, by Lemma 4.12,

$$(E\|\tilde{H}\|^2)^{1/2} \le \sqrt{2}(E\|H\|^2)^{1/2}$$

so it suffices to prove (4.15) with $\tilde{H}$ replacing H. Observe that for any real number α, $J(K, \alpha\sigma) = \alpha J(K, \sigma)$ i.e. that $J(K, \sigma)$ is a homogeneous function of the coefficients $\{a_\gamma | \gamma \in A\}$. Assume that there is no constant β_K for which (4.15) holds. We will reach a contradiction.

By the homogenity property of $J(K, \sigma)$ for each integer N we can find a process

$$\tilde{H}^N(x) = \sum_{\gamma \in \Gamma} a_\gamma^N \tilde{g}_\gamma \gamma(x), \quad x \in K$$

satisfying

$$E\|\tilde{H}^N(x)\|^2 \leq 2^{-N} \quad \text{and} \quad J(K, \sigma^N) \geq 2^N$$

where

$$\sigma^N(t-s) = (E|\tilde{H}^N(t) - \tilde{H}^N(s)|^2)^{1/2}$$

for all $s, t \in K$. We take the processes $\tilde{H}^N$ to be independent of each other and consider the process $\sum_{N=1}^{\infty} \tilde{H}^N$. Clearly this process has the same distribution as

$$\tilde{H}(x) = \sum a_\gamma \tilde{g}_\gamma \gamma(x), \quad x \in K$$

where $a_\gamma = \left(\sum_{N=1}^{\infty} |a_\gamma^N|^2\right)^{1/2}$. Since $\sum_{N=1}^{\infty} \tilde{H}^N$ is convergent in $L^2(C(K))$ the process $\tilde{H}$ has a version with continuous sample paths. Therefore, by Remark 4.13, so does H and by Theorem 3.2, $J(K, \sigma) < \infty$ with

$$\sigma(t-s) = (E|\tilde{H}(t) - \tilde{H}(s)|^2)^{1/2} = \left(\sum_{N=1}^{\infty} \sigma^N(t-s)^2\right)^{1/2}.$$

However since $\sigma \geq \sigma^N$ for all N, $J(K, \sigma^N) \geq 2^N$ for all N. Thus we establish a contradiction.

REMARK 4.16. Although we do not use this in the sequel, it is worthwhile to notice that it follows from Remark 4.13 that Theorem 3.2 and Corollary 4.15 are valid more generally when $H = \{H(t), t \in K\}$ is any complex valued process such that both Re H and Im H are real Gaussian processes and the metric $d(s,t) = (E|H(s)-H(t)|^2)^{1/2}$ is invariant under translation.

CHAPTER III

RANDOM FOURIER SERIES ON LOCALLY COMPACT ABELIAN GROUPS

1. *Continuity of random Fourier series*

Proof of Theorem 1.1.1. These results are trivial if G is discrete. Assume G is not discrete.

We first obtain the implications of $I(\sigma) < \infty$. Let $(\Omega_1, \mathcal{F}_1, P_1)$ denote the probability space of $\{\xi_\gamma\}$ and $(\Omega_2, \mathcal{F}_2, P_2)$ denote the probability space of $\{\varepsilon_\gamma\}$ and denote the corresponding expectation operators by E_1 and E_2. The series (1.1.11) is defined on the probability space $(\Omega_1 \times \Omega_2, \mathcal{F}_1 \times \mathcal{F}_2, P_1 \times P_2)$. We shall refer to this space as $(\Omega, \mathcal{F}, P)$ and denote the corresponding expectation operator by E.

Without loss of generality we can assume $\sup_{\gamma \epsilon A} E|\xi_\gamma|^2 \le 1$; the second assumption of (1.1.10) is not used in this part of the proof. Fix $\omega_1 \epsilon \Omega_1$ and consider

$$Z(x, \omega_1) = \sum_{\gamma \epsilon A} a_\gamma \varepsilon_\gamma \xi_\gamma(\omega_1) \gamma(x), \qquad x \epsilon K \tag{1.1}$$

as a random series on $(\Omega_2, \mathcal{F}_2, P_2)$. Note that $Z_1(x, \omega_1) = \sum_{\gamma \epsilon A} \varepsilon_\gamma \operatorname{Re}[a_\gamma \xi_\gamma(\omega_1)\gamma(x)]$ and $Z_2(x, \omega_1) = \sum_{\gamma \epsilon A} \varepsilon_\gamma \operatorname{Im}[a_\gamma \xi_\gamma(\omega_1)\gamma(x)]$ are both processes with sub-Gaussian increments with $\delta = 1$ (see e.g. Chapter 2, Section 5 [25]) and both $(E_2|Z_1(x,\omega_1) - Z_1(y,\omega_1)|^2)^{1/2}$ and $(E_2|Z_2(x,\omega_1) - Z_2(y,\omega_1)|^2)^{1/2}$ are less than or equal to

$$\sigma(x-y, \omega_1) = \Big(\sum_{\gamma \epsilon A} |a_\gamma|^2 |\xi_\gamma(\omega_1)|^2 |\gamma(x-y) - 1|^2\Big)^{1/2}. \tag{1.2}$$

By Theorem 2.3.1 with $t_0 = 0$ and Lemma 2.3.6 we have that whenever $I(\sigma(u,\omega_1)) < \infty$ there exists a version $\tilde{Z}(x,\omega_1)$ of $Z(x,\omega_1)$ such that

$$(1.3)\quad \{E_2[\sup_{x \epsilon K} |\tilde{Z}(x,\omega_1)|^2]\}^{1/2} \le D\Big[\Big(\sum_{\gamma \epsilon A} |a_\gamma|^2 |\xi_\gamma(\omega_1)|^2\Big)^{1/2} + I(\sigma(u,\omega_1))\Big],$$

for some constant D, where we also use the facts that

$$(1.4)\quad \bar{\sigma}(\omega_1) = \sup_{x \epsilon K \oplus K} \sigma(x,\omega_1) \le 2\Big(\sum_{\gamma \epsilon A} |a_\gamma|^2 |\xi_\gamma(\omega_1)|^2\Big)^{1/2}$$

and

$$(E_2|Z(0,\omega_1)|^2)^{1/2} = \Big(\sum_{\gamma \epsilon A} |a_\gamma|^2 |\xi_\gamma(\omega_1)|^2\Big)^{1/2}.$$

Furthermore, whenever $I(\sigma(u,\omega_1)) < \infty$, we have by Remark 2.4.4 that $Z(x,\omega_1)$ converges uniformly a.s. with respect to $(\Omega_2, \mathcal{F}_2, P_2)$. Therefore we can replace $\tilde{Z}(x,\omega_1)$ by $Z(x,\omega_1)$ in (1.3).

By (1.3) with Z replacing $\tilde{Z}$ we have

$$(1.5)\quad E_2 \sup_{x \epsilon K} |Z(x,\omega_1)|^2 \le D'\Big[\sum_{\gamma \epsilon A} |a_\gamma|^2 |\xi_\gamma(\omega_1)|^2 + |I(\sigma(u,\omega_1))|^2\Big],$$

where D' is an absolute constant. Next we apply E_1 to both sides of (1.5) and then take square roots to obtain

$$(1.6)\quad (E \sup_{x \epsilon K} |Z(x)|^2)^{1/2} \le D'\Big[\Big(\sum_{\gamma \epsilon A} |a_\gamma|^2\Big)^{1/2} + (E_1|I(\sigma(u,\omega_1))|^2)^{1/2}\Big].$$

where $Z(x)$ is given in (1.1.11). We now use Lemma 2.2.3 with $\|\ \ \|_\Omega = (E_1|\ \ |^2)^{1/2}$ to obtain

$$(1.7)\quad (E_1|I(\sigma(u,\omega_1)|^2)^{1/2} \le \int_0^{\mu_2} \frac{\overline{(E_1|\sigma(u,\omega_1)|^2)^{1/2}}}{u\left(\log \frac{4\mu_4}{u}\right)^{1/2}}\, du \le I(\sigma)$$

where σ is given in (1.1.12). Using (1.7) in (1.6) we obtain (1.1.15).

It is easy to see that the series (1.1.11) converges uniformly a.s. with respect to $(\Omega, \mathcal{F}, P)$. By (1.7) $I(\sigma) < \infty$ implies $I(\sigma(u, \omega_1)) < \infty$ a.s. with respect to $(\Omega_1, \mathcal{F}_1, P_1)$. Therefore, $Z(x, \omega_1)$ converges uniformly a.s. with respect to $(\Omega_2, \mathcal{F}_2, P_2)$ on a set $\overline{\Omega}_1 \subset \Omega_1$ with $P_1(\overline{\Omega}_1) = 1$. This implies, by Fubini's theorem, that the series (1.1.11) converges uniformly a.s.

We now obtain the implications of $I(\sigma) = \infty$. The main result in this direction is given in the following crucial lemma:

LEMMA 1.1. *There exist constants* C', C'' *depending only on* K *such that for any finite subset* F *of* A, *we have*

$$(1.8) \qquad \Big(E\Big\|\sum_{\gamma\epsilon F} a_\gamma \varepsilon_\gamma \xi_\gamma \gamma\Big\|^2\Big)^{1/2} \le C' \sup_{\gamma\epsilon F} (E|\xi_\gamma|^2)^{1/2} \cdot \Big(E\Big\|\sum_{\gamma\epsilon F} a_\gamma g_\gamma \gamma\Big\|^2\Big)^{1/2}$$

and

$$(1.9) \qquad \Big(E\Big\|\sum_{\gamma\epsilon F} a_\gamma g_\gamma \gamma\Big\|^2\Big)^{1/2} \le C''\Big(E\Big\|\sum_{\gamma\epsilon F} a_\gamma \varepsilon_\gamma \gamma\Big\|^2\Big)^{1/2}.$$

In particular, if $\sup E|\xi_\gamma|^2 < \infty$ *and* $\sum_{\gamma\epsilon A} a_\gamma \varepsilon_\gamma \gamma(x)$ *converges uniformly a.s. in some ordering, then so do* $\sum_{\gamma\epsilon A} a_\gamma g_\gamma \gamma(x)$ *and* $\sum_{\gamma\epsilon A} a_\gamma \varepsilon_\gamma \xi_\gamma \gamma(x)$ *and moreover (1.8) and (1.9) are true with* F *replaced by* A.

Proof. Consider $\sigma_F(t) = \Big(\sum_{\gamma\epsilon F} |a_\gamma|^2 |\gamma(t)-1|^2\Big)^{1/2}$. By Corollary 2.4.15 and (2.3.30) we have

$$(1.11) \qquad \Big(\Big(\sum_{\gamma\epsilon F} |a_\gamma|^2\Big)^{1/2} + I(\sigma)\Big) \le (1 + 2C_1 + 2\beta_K)\Big(E\Big\|\sum_{\gamma\epsilon F} a_\gamma g_\gamma \gamma\Big\|^2\Big)^{1/2}.$$

Therefore (1.8) follows immediately from (1.1.15) with $C' = C(1 + 2C_1 + 2\beta_K)$. To obtain (1.9) we write $g_\gamma = g'_\gamma + g''_\gamma$ where $g'_\gamma = g_\gamma I_{\{|g_\gamma|<N\}}$ (I_E is the

indicator function of the set E) and N is chosen large enough so that $(E|g''_\gamma|^2)^{1/2} = (2C')^{-1}$. We have then that

$$(1.12)\quad \Big(E\|\sum_F a_\gamma g_\gamma \gamma\|^2\Big)^{1/2} \le \Big(E\|\sum_F a_\gamma g'_\gamma \gamma\|^2\Big)^{1/2} + \Big(E\|\sum_F a_\gamma g''_\gamma \gamma\|^2\Big)^{1/2}.$$

By (2.4.6) and Remark 2.4.10 since $\{g'_\gamma\}$ is sign invariant, we obtain

$$\Big(E\|\sum_F a_\gamma g'_\gamma \gamma\|^2\Big)^{1/2} \le N\Big(E\|\sum_F a_\gamma \varepsilon_\gamma \gamma\|^2\Big)^{1/2}$$

and by (1.8)

$$\Big(E\|\sum_F a_\gamma g''_\gamma \gamma\|^2\Big)^{1/2} \le \frac{1}{2}\Big(E\|\sum_F a_\gamma g_\gamma \gamma\|^2\Big)^{1/2}.$$

Combining this with (1.12) we obtain (1.9) with $C'' = 2N$. Finally, if $\sum_{\gamma\epsilon A} a_\gamma \varepsilon_\gamma \gamma$ converge uniformly a.s. in some ordering, then (cf. Theorem 2.4.5) it is convergent in $L^2(dP; C(K))$. Therefore by (1.9) and (1.8) the same holds for $\sum_{\gamma\epsilon A} a_\gamma g_\gamma \gamma$ and for $\sum_{\gamma\epsilon A} a_\gamma \varepsilon_\gamma \xi_\gamma \gamma$ and the final assertion is clear. (Recall that by Lemma 2.4.1 convergence in $L^2(C(K))$ implies a.s. convergence for series such as these.)

It follows from (1.1.15), Theorem 2.3.2, Theorem 2.4.3 and Lemma 1.1 that $I(\sigma) < \infty$ is a necessary and sufficient condition for the uniform convergence a.s. of $\sum a_\gamma \varepsilon_\gamma \gamma$. If we had restricted ourselves to sequences $\{\xi_\gamma\}$ that are real and independent, as well as satisfying (1.1.10) we could complete the proof of Theorem 1.1.1 (at least in the case with U replaced by K in (1.1.16)) by using known results (cf. [24]). What follows is necessitated by the greater generality of $\{\xi_k\}$.

LEMMA 1.2. *Using the notation of Theorem 1.1.1 with* $\liminf_{k\to\infty} E|\xi_k| > 0$ *and* $\sup_k E|\xi_k|^2 = 1$, *let* $\{\gamma_k\}$ *be some ordering of* $A \subset \Gamma$. *Then if*

(1.13) $$M = \sup \left\| \sum_{k=0}^{n} a_k \varepsilon_k \xi_k \gamma_k \right\| < \infty \ a.s.$$

it follows that

(1.14) $$EM^2 < \infty .$$

Proof. Without loss of generality we can assume that $\inf_k E|\xi_k| > 0$. Let

$$M_j = \sup_{n \leq j} \left\| \sum_{k=0}^{n} a_k \varepsilon_k \xi_k \gamma_k \right\| .$$

We use the notation introduced at the beginning of this section. By Theorem 2.4.9 and (2.4.2′) we have, for all j,

(1.15) $$EM_j \geq \frac{1}{4} \inf_k E|\xi_k| E \sup_{n \leq j} \left\| \sum_{k=0}^{n} a_k \varepsilon_k \gamma_k \right\| .$$

On the other hand by (1.8), (1.9) and (2.4.2′) we have

(1.16) $$(EM_j^2)^{1/2} \leq C_1 \left(E \sup_{n \leq j} \left\| \sum_{k=0}^{n} a_k \varepsilon_k \gamma_k \right\|^2 \right)^{1/2}$$

for some constant C_1 independent of $\{a_k\}$ (and hence of j). Let $\inf_k E|\xi_k| = 2\alpha$. Applying Corollary 2.4.6 with $q = 1$ and $p = 2$ to (1.15) and using (1.16) we get

(1.17) $$(EM_j^2)^{1/2} \leq (D/\alpha) EM_j$$

for some constant D independent of $\{a_k\}$. By (1.17) and a simple inequality of Paley and Zygmund ([26], page 6) there exists a $\delta > 0$ such that

$$P(M_j > \delta (EM_j^2)^{1/2}) > \delta .$$

Also by (1.13) we know that there exists a $\beta > 0$ such that

$$P(M_j > \beta) < \delta/2$$

for all j. Therefore $(EM_j^2)^{1/2} < \beta/\delta$ and (1.14) follows from the monotone convergence theorem.

We now complete the proof of Theorem 1.1.1. Combining (1.15) with Corollary 2.4.6 (with $q=1$ and $p=2$) and (1.9) we get

$$\text{(1.18)} \qquad E \sup_n \left\| \sum_{k=0}^{n} a_k \varepsilon_k \xi_k \gamma_k \right\| \geq d\left(E \sup_n \left\| \sum_{k=0}^{n} a_k g_k \gamma_k \right\|^2\right)^{1/2}$$

for some constant $d > 0$ independent of $\{a_k\}$. If $I(\sigma) = \infty$ the term on the right is infinite by Theorem 2.3.6. Consequently, by Lemma 1.2, $M = \sup_n \left\| \sum_{k=0}^{n} a_k \varepsilon_k \xi_k \gamma_k \right\|$ is unbounded on a set of positive measure. This completes the proof of Theorem 1.1 with U replaced by K.

Now let $U \subset K$ be open. If $I(\sigma) = \infty$ we can find (by Theorem 2.3.7) a compact symmetric set $K' \subset U$ such that $I(K', \sigma) < \infty$ remains a necessary and sufficient condition for

$$\sup_n \sup_{x \epsilon K'} \left| \sum_{k=0}^{n} a_k g_k \gamma_k \right| < \infty \text{ a.s.}$$

We can now obtain Lemma's 1.1 and 1.2 for the norm in the space $C(K')$ and in particular we can obtain (1.18) for this norm. Thus $I(\sigma) = \infty$ implies

$$\sup_n \sup_{x \epsilon K'} \left| \sum_{k=0}^{n} a_k \varepsilon_k \xi_k \gamma_k \right| = \infty$$

on a set with probability greater than zero.

REMARK 1.3. If $\{\xi_k\}$ is a sequence of independent random variables the event (1.1.16) is a tail event and occurs either with probability zero or 1. However, for $\{\xi_k\}$ sign-invariant this need not be the case. For example we can choose $\{\xi_k\}$ as follows: Let the probability space be $[0,1]$ with Lebesgue measure, $\omega \in [0,1]$, and let

$$\xi_k(\omega) = \begin{cases} 0 & 0 \le \omega < 1/2 \\ \varepsilon_k(\omega) & 1/2 \le \omega \le 1 \end{cases}$$

where $\varepsilon_k(\omega)$ is the k-th classical Rademacher function on $[0,1]$. There are many examples of groups G and sequences $\{\gamma_k\}$ for which there exists a sequence $\{a_k\} = \ell^2$ such that

$$\sup_n \|\sum_{k=0}^{n} a_k \xi_k \gamma_k\| = 0$$

with probability 1/2 and is infinite with probability 1/2. Also in the case $I(\sigma) = \infty$ we cannot remove the condition $\liminf_{k\to\infty} E|\xi_k| > 0$ since without this condition we can find a subsequence $\{\xi_{k_j}\}$ of $\{\xi_k\}$ such that

$$(1.19) \qquad \sum_{j=1}^{\infty} P(|\xi_{k_j}| > 2^{-j}) < \infty .$$

If (1.19) holds one can readily find groups G and sequences $\{\gamma_k\}$ and $\{a_k\} \in \ell^2$ for which $\sum_{k=0}^{\infty} a_k \varepsilon_k \xi_k \gamma_k$ converges uniformly a.s. and for which $I(\sigma) = \infty$.

The proof of Corollary 1.1.2 is simple. Consider

$$\tilde{Z}(x) - \tilde{Z}'(x) = \sum_{\gamma \in A} a_\gamma(\xi_\gamma - \xi'_\gamma)\gamma(x), \qquad x \in K$$

where $\{\xi'_\gamma\}$ is an independent copy of $\{\xi_\gamma\}$. The sequence $\{\xi_\gamma - \xi'_\gamma\}$ is sign-invariant and satisfies (1.10). (Note that $E|\xi_\gamma - \xi'_\gamma| \ge E|\xi_\gamma|$ follows from the proof of Lemma 2.4.2.) Therefore Theorem 1.1 holds for $\tilde{Z}(x) - \tilde{Z}'(x)$ and Lemma 2.4.2 gives Corollary 1.2.

We now obtain a result analagous to (2.3.4) but for the random Fourier series $Z(x)$ as given in (1.1.11). Recall that the compact set $K \subset G$ is metrizable, we will denote its metric by $\tau(x, y)$, $x, y \in K$. Otherwise we continue to use the notation of this section.

THEOREM 1.4. *For* $Z(x)$ *as given in (1.1.11) we have*

(1.20)
$$E\Big[\sup_{\substack{\tau(x,y)\leq a \\ x,y\in K}} |Z(x)-Z(y)|\Big] \leq C\big(\sup_{\gamma} E|\xi_\gamma|^2\big)^{1/2} \cdot \left[\int_0^{g(a)} (\log N_\sigma(K,u))^{1/2}\,du + g(a)h(1/4)\right]$$

where $h(d) = d(\log\log 1/d)^{1/2}$ *and* $\lim_{a\to 0} g(a) = 0$, *and* C *is a constant.*

Proof. Let $Z(x, \omega_1)$ and $\sigma(x-y, \omega_1)$ be as in (1.1) and (1.2). Define $\hat{\sigma}(a, \omega_1) = \sup_{\tau(x,y)\leq a} \sigma(x-y, \omega_1)$ and note that $\tau(x,y) \leq a$ implies $\sigma(x-y, \omega_1) \leq \hat{\sigma}(a, \omega_1)$. By (2.3.4) we have

(1.21)
$$E_2\Big[\sup_{\substack{\tau(x,y)\leq a \\ x,y\in K}} |Z(x,\omega_1)-Z(y,\omega_1)|\Big]$$
$$\leq E_2\Big[\sup_{\substack{\sigma(x-y,\omega_1)\leq\hat{\sigma}(a,\omega_1) \\ x,y\in K}} |Z(x,\omega_1)-Z(y,\omega_1)|\Big]$$
$$\leq K'\left[\int_0^{\hat{\sigma}(a,\omega_1)} (\log N_{\sigma(\cdot,\omega_1)}(K,u))^{1/2}\,du + 4\hat{\sigma}(a,\omega_1)h(1/4)\right],$$

for some constant K'. We might as well assume that $\sup_{\gamma} E|\xi_\gamma|^2 \leq 1$; (1.20) will then follow by a renormalization. We define

$$g(a) = E_1\hat{\sigma}(a, \omega_1).$$

It is easy to see that $\lim_{a\to 0} g(a) = 0$. Let γ_n be an ordering of $\{\gamma | \gamma \in A\}$ and for simplicity denote $a_{\gamma_n} = a_n$, $\varepsilon_{\gamma_n} = \varepsilon_n$ and $\xi_{\gamma_n} = \xi_n$ so that

$$\hat{\sigma}(a, \omega_1) = \Big(\sup_{\tau(u,0)\le a} \sum_{n=1}^{\infty} |a_n|^2 |\xi_n(\omega_1)|^2 |\gamma_n(u)-1|^2 \Big)^{1/2} .$$

Given $\varepsilon > 0$, choose N such that $\sqrt{2}\Big(\sum_{n>N} |a_n|^2\Big) < \varepsilon/2$. Choose a_ε such that $\sup_{\tau(0,u)\le a_\varepsilon} |\gamma_n(u)-1| < \varepsilon/\Big(\sum_1^\infty |a_n|^2\Big)^{1/2}$ for $1 \le n \le N$. Then

$$\hat{\sigma}(a_\varepsilon, \omega_1) \le \Big(\sup_{\tau(0,u)\le a_\varepsilon} \sum_1^N |a_n|^2 |\xi_n(\omega_1)|^2 |\gamma_n(u)-1|^2 \Big)^{1/2}$$

$$+ \Big(2 \sum_{n>N} |a_n|^2 |\xi_n(\omega_1)|^2\Big)^{1/2}$$

and $E_1(\hat{\sigma}(a_\varepsilon, \omega_1)) \le \varepsilon$ so $\lim_{a\to 0} g(a) = 0$. We complete the proof by applying E_1 to each side of (1.21) and using Lemma 2.2.4 with $\phi(t) = (\log(1+1/t))^{1/2}$. (This approach is used in a similar way in [10].) By (2.1.9) and (2.1.2) we have

(1.22) $$N_{\sigma(\cdot,\omega_1)}(K, u) \le N_{\sigma(\cdot,\omega_1)}(K \oplus K, u/2) \le \frac{\mu_4}{m_{\sigma(\cdot,\omega_1)}(u/4)}$$

which is, by definition,

$$= \frac{\mu_4}{\mu\{x \in K \oplus K \,|\, \sigma(x,\omega_1) < u/4\}} .$$

By (1.22)

$$(\log N_{\sigma(\cdot,\omega_1)}(K, u))^{1/2} \le \Big(\log\Big(1 + \frac{\mu_4}{\mu\{x \in K \oplus K \,|\, \sigma(x,\omega_1) < u/4\}}\Big)\Big)^{1/2} .$$

Therefore, by Lemma 2.2.4 we have

$$\text{(1.23)} \qquad E_1 \int_0^{\hat{\sigma}(a,\omega_1)} (\log N_{\sigma(\cdot,\omega_1)}(K,u))^{1/2} du$$

$$\leq \int_0^{g(a)} \left(\log\left(1 + \frac{\mu_4}{\mu\{x \in K \oplus K \mid \sigma(x) < u/4\}}\right)\right)^{1/2} du$$

$$= \int_0^{g(a)} \left(\log\left(1 + \frac{\mu_4}{m_\sigma(u/4)}\right)\right)^{1/2} du \ .$$

Therefore, as we stated, we apply E_1 to each side of (1.21) and use (1.23) to get (1.20). (We also need (2.1.3) to replace the last term in (1.23) by $\int_0^{g(a)} (\log N_\sigma(K,u))^{1/2} du$.)

2. *Random Fourier series on the real line*

Proof of Theorem 1.1.3. Following the remarks preceding Theorem 1.1.3 we need only show that (1.1.20) implies (1.1.19). Suppose

$$\text{(2.1)} \qquad \sup_n \sup_{t \in I} \left|\sum_{k=0}^n a_k \varepsilon_k \rho_k \cos(\lambda_k t + \Phi_k)\right| < \infty \text{ a.s.};$$

then in order for (1.1.20) to hold there must exist a $t_0 \in I$ and a $\delta > 0$ such that $\{t : |t-t_0| < \delta\} \subset I$ and

$$\text{(2.2)} \qquad \sup_n \sup_{|t-t_0|<\delta/2} \left|\sum_{k=0}^n a_k \varepsilon_k \rho_k \sin(\lambda_k t + \Phi_k)\right| = \infty$$

on a set of probability greater than zero. However, we will show that (2.1) implies that the event in (2.2) is finite a.s.

Let $Y_n(t) = \sum_{k=0}^{n} a_k \varepsilon_k \rho_k \cos(\lambda_k t + \Phi_k)$. Assuming (2.1) we have

$\sup_n \sup_{|t-t_0|\leq\delta} |Y_n(t)| < \infty$ a.s. which implies

$$(2.3) \qquad \sup_n \sup_{\substack{|t-t_0|\leq\delta/2 \\ |s|\leq\delta/2}} |Y_n(t+s)-Y_n(t-s)| < \infty \text{ a.s.}$$

We use the notation of Chapter I for the product probability space of $\{\xi_k\}$ and $\{\varepsilon_k\}$ where we take $\xi_k = \rho_k e^{i\Phi_k}$. It follows from (2.3) that there exists a set $\overline{\Omega_1} \subset \Omega_1$, $P_1(\overline{\Omega_1}) = 1$ such that for $\omega_1 \in \overline{\Omega_1}$

$$(2.4) \quad \sup \sup_{\substack{|t-t_0|\leq\delta/2 \\ |s|\leq\delta/2}} \left|\sum_{k=0}^{n} a_k \varepsilon_k \rho_k(\omega_1) \sin(\lambda_k t + \Phi_k(\omega_1)) \sin \lambda_k s\right| < \infty \text{ a.s. } (P_2).$$

Since for each $\omega_1 \in \overline{\Omega}_1$ the series in (2.4) is a Rademacher series we have by Theorem 2.4.5 that for each $\omega_1 \in \overline{\Omega}_1$

$$E_2\left\{\sup_n \sup_{\substack{|t-t_0|\leq\delta/2 \\ |s|\leq\delta/2}} \left|\sum_{k=0}^{n} a_k \varepsilon_k \rho_k(\omega_1) \sin(\lambda_k t + \Phi_k(\omega_1)) \sin \lambda_k s\right|\right\} < \infty .$$

Therefore

$$\int_0^{\delta/2} E_2\left\{\sup_n \sup_{|t-t_0|\leq\delta/2} \left|\sum_{k=0}^{n} a_k \varepsilon_k \rho_k(\omega_1) \sin(\lambda_k t + \Phi_k(\omega_1)) \sin \lambda_k s\right|\right\} ds < \infty .$$

By (2.4.6) and (2.4.2′) we can replace $\sin \lambda_k s$ by $|\sin \lambda_k s|$. Thus we obtain

$$E_2\left\{\sup_n \sup_{|t-t_0|\leq\delta/2} \left|\sum_{k=0}^{n} a_k \varepsilon_k \rho_k(\omega_1) \sin(\lambda_k t + \Phi_k(\omega_1)) \int_0^{\delta/2} |\sin \lambda_k s|\, ds\right|\right\} < \infty .$$

Let us assume that $\inf_k |\lambda_k| > 1$. Then

$$\inf_k \int_0^{\delta/2} |\sin \lambda_k s| ds > 0$$

and we conclude that for each $\omega_1 \in \overline{\Omega}_1$

$$E_2 \sup_n \sup_{|t-t_0|\le\delta/2} \left|\sum_{k=0}^{n} a_k \varepsilon_k \rho_k(\omega_1) \sin(\lambda_k t + \Phi_k(\omega_1))\right| < \infty$$

which implies that

(2.5) $$\sup_n \sup_{|t-t_0|\le\delta/2} \left|\sum_{k=0}^{n} a_k \varepsilon_k \rho_k \sin(\lambda_k t + \Phi_k)\right| < \infty \text{ a.s. (P)}.$$

Now we remove the restriction $\inf |\lambda_k| > 1$. Define $J = \{k : |\lambda_k| \le 1\}$ and $K = \{k : |\lambda_k| > 1\}$. From (2.5) we have

(2.6) $$\sup_n \sup_{|t-t_0|\le\delta/2} \left|\sum_{\substack{k=1\\k\in K}}^{n} a_k \varepsilon_k \rho_k \sin(\lambda_k t + \Phi_k)\right| < \infty \text{ a.s.}$$

By Theorem 1.1.1 we have

$$\sup_{t\in[0,1]} \left|\sum_{k\in J} a_k \varepsilon_k \rho_k \sin(\lambda_k t + \Phi_k)\right| < \infty \text{ a.s.}$$

since

$$\overline{\sigma_J(h)} = \overline{\left(\sum_{k\in J} a_k^2 \sin^2 \frac{\lambda_k h}{2}\right)^{1/2}} \le \left(\sum a_k^2\right)^{1/2} h$$

and consequently $I(\sigma_J) < \infty$. By Lemma 2.4.1

(2.7) $$\sup_n \sup_{t\in[0,1]} \left|\sum_{\substack{k=1\\k\in J}}^{n} a_k \varepsilon_k \rho_k \sin(\lambda_k t + \Phi_k)\right| < \infty \text{ a.s.}$$

Combining (2.6) and (2.7) we obtain a contradiction to (2.2).

3. *Random Fourier series on compact Abelian groups*

Proof of Theorem 1.1.4. Assume first that G is not discrete. By (1.3) we have

$$(3.1)\qquad E_2 \sup_{x\epsilon G} |Z(x,\omega_1)| \le D\Big[\Big(\sum_{\gamma\epsilon A} |a_\gamma|^2 |\xi_\gamma(\omega_1)|^2\Big)^{1/2} + I(\sigma(u,\omega_1))\Big]$$

where D is a numerical constant. As we observed in (1.7) we also have

$$(3.2)\qquad E_1(I(\sigma(u,\omega_1))) \le I(\sigma)\,.$$

Therefore applying E_1 to both sides of (3.1) and using (3.2) we get the right side of (1.1.21).

By Theorem 1.1.1 all three phrases in the inequality (1.1.21) are infinite if $I(\sigma) = \infty$. Therefore we need only consider the case $I(\sigma) < \infty$. In this case, again by Theorem 1.1.1, all the series we consider below converge uniformly a.s. By Theorem 2.3.4 and (2.3.30) we have

$$(3.3)\qquad D_1\Big[\Big(\sum_{\gamma\epsilon A} |a_\gamma|^2\Big)^{1/2} + I(\sigma)\Big] \le E \sup_{x\epsilon G} \Big|\sum_{\gamma\epsilon A} a_\gamma g_\gamma \gamma(x)\Big|$$

for some numerical constant $D_1 > 0$. Combining (3.3) and the right side of (1.1.21) we have

$$(3.4)\qquad E \sup_{x\epsilon G} \Big|\sum_{\gamma\epsilon A} a_\gamma \varepsilon_\gamma \xi_\gamma \gamma\Big| \le D_2(\sup_{\gamma\epsilon A} E|\xi_\gamma|^2)^{1/2} E \sup_{x\epsilon G} \Big|\sum_{\gamma\epsilon A} a_\gamma g_\gamma \gamma(x)\Big|\,.$$

Thus we obtain an expression similar to (1.8) without using Corollary 2.4.15, and D_2 is a numerical constant. Following the proof of (1.8) im-(1.9) in Lemma 1.1 we have

$$(3.5)\qquad E \sup_{x\epsilon G} \Big|\sum_{\gamma\epsilon A} a_\gamma g_\gamma \gamma(x)\Big| \le D_3 E \sup_{x\epsilon G} \Big|\sum_{\gamma\epsilon A} a_\gamma \varepsilon_\gamma \gamma(x)\Big|$$

where D_3 is a numerical constant.

By (2.4.7), we have (with $\|\cdot\|$ the norm in $C(G)$),

$$E\|\sum a_\gamma \varepsilon_\gamma \xi_\gamma\| \geq \frac{1}{2}(\inf_{\gamma \epsilon A} E|\xi_\gamma|)E\|\sum_{\gamma \epsilon A} a_\gamma \varepsilon_\gamma \gamma\| . \tag{3.6}$$

We now combine (3.6), (3.5) and (3.3) to obtain the left side of (1.1.21), and this completes the proof in the non-discrete case. Contrary to the preceding results, Theorem 1.1.4 does not become trivial if G is assumed to be a finite discrete group, because the constants in (1.1.21) are now absolute constants, independent in particular of the cardinality of G. However, if G is a finite group, then we can select any compact non-discrete group – for instance the circle group T –and work with $G \times T$ in the place of G. Clearly, a random Fourier series on G can be identified with a random Fourier series on $G \times T$ independent of the second coordinate. Since $G \times T$ is non-discrete, this substitution allows us to establish (1.1.21) also when G is a finite group.

CHAPTER IV

THE CENTRAL LIMIT THEOREM AND RELATED QUESTIONS

In this chapter, we prove that random Fourier series of the form (1.1.11) or its variant (1.1.17) satisfy the Central Limit Theorem (in short, the CLT) on C(K) as soon as they have a version with continuous sample paths. This is not surprising since in [36] it was proved (for $\Gamma = \mathbf{R}$) that these series satisfy the CLT as soon as they are pre-Gaussian, and the results in Chapter III show that, if they are continuous, they are "automatically" pre-Gaussian. We first recall some basic facts about the CLT for Banach space valued random variables.

Let X be a random variable with values in a separable Banach space B. Let μ be the probability measure induced on B by X and assume that X has zero mean, that is, $\int_B x\mu(dx) \equiv E(X) = 0$. The covariance of X, or equivalently of μ, is a function on $B^* \times B^*$ where B^* is the space of all R-linear forms on B given by

$$\Gamma_\mu(f,g) = \int_B f(x)\,g(x)\,\mu(dx), \qquad f,g \in B^* .$$

The random variable X is called pre-Gaussian if there exists a Gaussian measure ν on B such that for all $f,g \in B^*$

$$\Gamma_\mu(f,g) = \Gamma_\nu(f,g) . \tag{1.1}$$

The random variable X, as defined, is said to satisfy the CLT on B if

$$\lim_{n\to\infty} Ef\left(\frac{X_1 + \cdots + X_n}{n^{1/2}}\right) = Ef(Y) \tag{1.2}$$

for all bounded real valued continuous functions f on B, where $X_1, X_2, \cdots$ are independent copies of X and Y is a B valued Gaussian random variable (with corresponding measure ν) that satisfies (1.1). By considering restrictions of X to finite dimensional subspaces of B one can see that if the left side of (1.2) has a limit then the limit must be $Ef(Z)$. Therefore a necessary condition for X to satisfy the Central Limit Theorem on B is that it be pre-Gaussian.

We now define $\{\check{g}_\gamma\}$ to be independent complex valued Gaussian random variables (i.e. both the real and imaginary part of $\check{g}_\gamma$ are Gaussian), satisfying $E\check{g}_\gamma = 0$ and $E|\check{g}_\gamma|^2 = 1$. By Theorem 1.1.1, $I(\sigma) < \infty$ if and only if $\sum a_\gamma \check{g}_\gamma \gamma(x)$, $x \epsilon K$ convergences uniformly a.s. (or equivalently has a version with continuous sample paths).

In what follows we consider the series $Z(x)$ given in (1.1.11) with $E|\xi_\gamma|^2 = 1$ for all γ. (We do not require $\inf E|\xi_\gamma| > 0$.) This formulation will impose no loss of generality since given an arbitrary sequence of complex valued random variables $\{u_\gamma\}$ with $\sup_\gamma E|u_\gamma|^2 < \infty$ one can write (1.1.11) as

$$\sum_{\gamma \epsilon A} a_\gamma (E|u_\gamma|^2)^{1/2} \varepsilon_\gamma (u_\gamma/(E|u_\gamma|^2)^{1/2}) \gamma(x), \qquad x \epsilon K$$

and the theorem below will follow with $\{a_\gamma\}$ in (1.1.11) replaced by $\{a_\gamma (E|u_\gamma|^2)^{1/2}\}$. As we stated above if $Z(x)$ satisfies the CLT then the process to which it converges is the Gaussian process $\sum_{\gamma \epsilon A} a_\gamma \check{g}_\gamma \gamma(x)$, $x \epsilon K$ where $E|\mathrm{Re}\, \check{g}_\gamma|^2 = E|\mathrm{Re}\, \xi_\gamma|^2$, $E|\mathrm{Im}\, \check{g}_\gamma|^2 = E|\mathrm{Im}\, \xi_\gamma|^2$ and $E(\mathrm{Re}\, \check{g}_\gamma)(\mathrm{Im}\, \check{g}_\gamma) = E(\mathrm{Re}\, \xi_\gamma)(\mathrm{Im}\, \xi_\gamma)$. Therefore a necessary condition for $Z(x)$ to satisfy the CLT is that $I(\sigma) < \infty$.

THEOREM 1.1. *In the above notation with* $E|\xi_\gamma|^2 = 1$ *we have that the* $C(K)$ *valued random variable* Z *given in (1.1.11) satisfies the CLT if and only if* $I(\sigma) < \infty$. *(In fact, if* $I(\sigma) = \infty$, Z *is not even pre-Gaussian.)*

We will provide two proofs of this theorem.

Proof I. By the results of Chapter 3, if $I(\sigma) < \infty$ the series $Z(x)$ given in (1.1.11) converges in $L_2(C(K))$ and $E\|Z\|^2 < \infty$. Let $\{\gamma_n | n \in \mathbf{N}\}$ be an enumeration of the set A. We write simply $\varepsilon_n = \varepsilon_{\gamma_n}$, $\xi_n = \xi_{\gamma_n}$ and $a_n = a_{\gamma_n}$ so that

$$(1.3) \qquad Z = \sum_{n=0}^{\infty} a_n \varepsilon_n \xi_n \gamma_n .$$

For every integer N we write

$$Z^N = \sum_{n>N} a_n \varepsilon_n \xi_n \gamma_n ;$$

let $Z_1^N, Z_2^N, \cdots$ be a sequence of independent copies of the $C(K)$-valued random variable Z^N. It is well known, as we will recall below, that to check that Z satisfies the CLT on $C(K)$, it is sufficient to prove that

$$(1.4) \qquad \delta_N = \sup_{m \in \mathbf{N}} E\left\|\frac{Z_1^N + \cdots + Z_m^N}{\sqrt{m}}\right\|^2$$

is finite and tends to zero when $N \to \infty$. Let $\{\eta_n^1\}, \{\eta_n^2\}, \cdots,$ be a sequence of independent copies of the sequence $\{\varepsilon_n \xi_n\}$. Denote by $s_n(m)$ the variable $s_n(m) = 1/\sqrt{m} \sum_{j=1}^{m} \eta_n^j$.

We observe that $E|s_n(m)|^2 = 1$ for all $n, m \in \mathbf{N}$. Moreover, the sequence $\{s_n(m)\}$ is sign invariant. If $\{\theta_n\}$ is a Rademacher sequence (i.e. a copy of $\{\varepsilon_n\}$) independent of all the variables previously introduced, the distribution of $\{\theta_n s_n(m)\}_n$ is the same as that of $\{s_n(m)\}_n$. Therefore, we have

$$(1.5) \qquad E\left\|\sum_{j=1}^{m} Z_j^N/\sqrt{m}\right\|^2 = E\left\|\sum_{n>N} \theta_n s_n(m) a_n \gamma_n\right\|^2 .$$

By Theorem 1.1.1 applied to the right side of (1.4) we have that

$$E\Big\|\sum_{j=1}^{m} Z_j^N/\sqrt{m}\Big\|^2 \le C\lambda_N = C'\Big[\Big(\sum_{k=N}^{\infty} |a_n|^2\Big)^{1/2} + I(\sigma_N)\Big]$$

where $\sigma_N(u) = \Big(\sum_{n>N} |a_n|^2|\gamma_n(u)-1|^2\Big)^{1/2}$. Since $I(\sigma) < \infty$, by the dominated convergence theorem we see that $\lambda_N \to 0$ as $N \to \infty$. Hence $\delta_N \to 0$ as $N \to \infty$. The fact that (1.4) is sufficient to prove the CLT comes from the following lemma.

LEMMA 1.2. *Let* X *be a random variable with values in a separable Banach space* B. *Suppose that for each* $\varepsilon > 0$ *there exists a* B *valued random variable* Y_ε *such that*

(i) Y_ε *satisfies the CLT*

and

(ii) $E\|1/\sqrt{m}\sum_{i=1}^{m}(X_i - Y_{\varepsilon,i})\| \le \varepsilon$

where $\{X_i\}$ *are independent copies of* X *and* $\{Y_{\varepsilon,i}\}$ *are independent copies of* Y_ε.

Then X *satisfies the CLT.*

Proof of Lemma 1.2. This lemma is part of Theorem 3.1 [45]. In this theorem X is taken to be finite dimensional but the only property of X that is used is that it satisfies the CLT. This observation was also made in [10].

Completion of Proof I. Since $\sum_{n\le N} a_n\varepsilon_n\xi_n\gamma_n$ has finite dimensional range it satisfies the CLT by classical considerations. Also, Condition (ii) of Lemma 1.2 holds because $\delta_N \to 0$ as $N \to \infty$. This completes the proof when $I(\sigma) < \infty$ and we have already established that Z is not even pre-Gaussian when $I(\sigma) = \infty$.

Proof II. Let Z_i be independent copies of Z and define

$$S_m = 1/\sqrt{m} \sum_{i=1}^{m} Z_i = \sum_{n=1}^{\infty} a_n s_n(m) \gamma_n$$

for $s_n(m)$ as defined above. If $I(\sigma) < \infty$ we have by Theorem 3.1.4 and (2.3.30) that

$$E[\sup_{\substack{\tau(x,y)\leq a \\ x,y\in K}} |S_m(x) - S_m(y)|] \leq F(\sigma, a)$$

where τ is a metric on K and F is a function that depends on σ and a but not on m and which satisfies $\lim_{a\to 0} F(\sigma, a) = 0$. Therefore, the measures induced on $C(K)$ by the $\{S_m\}$ are tight (see e.g. [3]). It follows by the finite dimensional CLT that $\{S_m\}$ converges weakly to the appropriate Gaussian limit. The case $I(\sigma) = \infty$ follows as above.

REMARK 1.3. The same argument as above shows that the random series Q defined in (1.1.17) satisfies the CLT on $C([0,1])$ as soon as it is continuous a.s.

REMARK 1.4. Since $E\|Z\|^2 < \infty$, Theorem 4.3 in [45] shows that Z satisfies the Law of the Iterated Logarithm as soon as it satisfies the CLT.

The Banach spaces B which have the property that every pre-Gaussian B-valued random variable satisfies the CLT, have been characterized (see e.g. [22], Theorem 5); they are exactly the Banach spaces which are of cotype 2.

A Banach space B is said to be of cotype 2 if there exists a constant λ such that for all integers N and all sequences $\{x^1, \cdots, x^N\}$ of elements of B we have

$$\left(\sum_1^N \|x^j\|^2\right)^{1/2} \leq \lambda E\|\sum_1^N g_j x^j\|$$

where $\{g_n\}$, as above, is a sequence of independent $N(0,1)$ Gaussian random variables. (For more information on this notion see, e.g. [42].)

The preceding Theorem 1.1 suggests that there is a space of cotype 2 connected with these questions. Clearly $C(K)$ itself is not of cotype 2 unless K is a finite set; however, the space B defined below is of cotype 2. We define the elements of B as the families $a = \{a_\gamma | \gamma \in \Gamma\}$ of complex numbers such that $\sum_\Gamma |a_\gamma|^2 < \infty$ and such that the series $\sum_{\gamma \in \Gamma} a_\gamma \varepsilon_\gamma \gamma(x), x \in K$ is a.s. continuous on K (where K is, as before, a fixed compact neighborhood of 0 in G). The norm on B is defined by

$$|||a||| = \Big(E\|\sum_{\gamma \in \Gamma} a_\gamma \varepsilon_\gamma \gamma\|^2_{C(K)}\Big)^{\frac{1}{2}}.$$

It is easy to see that B is a Banach space and we can observe, using (2.4.9), that for all a in B

$$\frac{1}{2}|||a||| \le |||\{|a_\gamma| \mid \gamma \in \Gamma\}||| \le 2|||a||| . \tag{1.6}$$

THEOREM 1.5. *The Banach space* B *is of cotype* 2.

Proof. Consider $a^1, \cdots, a^N$ in B. Define the sequence $a = \{a_\gamma | \gamma \in \Gamma\}$ where $a_\gamma = \Big(\sum_{j=1}^N |a_\gamma^j|^2\Big)^{\frac{1}{2}}$. We first claim that

$$\Big(\sum_{j=1}^N |||a^j|||^2\Big)^{\frac{1}{2}} \le C'C''|||a||| \tag{1.7}$$

where C', C'' are the same as in Lemma 3.1.1. Indeed, let $A_1, \cdots, A_N$ be disjoint sets with $P(A_j) = 1/N$ for $j = 1, \cdots, N$. Consider random variables ξ_γ defined by $\xi_\gamma = \sum_{j=1}^N a_\gamma^j / a_\gamma I_{A_j} \sqrt{N}$. We may as well assume, as before, that $\{\varepsilon_\gamma\}$ and $\{\xi_\gamma\}$ are independent of each other. By (3.1.8) and (3.1.9) we have, since $E|\xi_\gamma|^2 = 1$,

$$E\|\sum a_\gamma \varepsilon_\gamma \xi_\gamma \gamma\|^2 \le (C'C'')^2 E\|\sum a_\gamma \varepsilon_\gamma \gamma\|^2$$

or equivalently

$$E|||\{a_\gamma \xi_\gamma\}|||^2 \le (C'C'')^2 |||a|||^2 , \tag{1.8}$$

and since we have obviously that

$$E|||\{a_\gamma \xi_\gamma\}|||^2 = \sum_{j=1}^{N} |||a^j|||^2 ,$$

we see that (1.8) implies (1.7). Theorem 1.4 follows from (1.7) and from the next inequality,

$$\sqrt{2/\pi}\,|||a||| \le 2E|||\sum_1^N g_j a^j||| . \tag{1.9}$$

Therefore, it remains to prove (1.9). We have, since $E|g_j| = \sqrt{2/\pi}$, that for all $\gamma \epsilon \Gamma$

$$\sqrt{2/\pi}\, a_\gamma = E|\sum_{j=1}^{N} g_j a_\gamma^j| .$$

Hence, by the convexity of the norm,

$$\begin{aligned}\sqrt{2/\pi}\,|||a||| &= |||\{E|\sum_1^N g_j a_\gamma^j|\}||| \\ &\le E|||\{|\sum_1^N g_j a_\gamma^j|\}||| \\ &\le 2E|||\sum_1^N g_j a^j||| .\end{aligned}$$

where at the last step we use (1.6). This proves (1.9) and completes the proof of Theorem 1.4.

REMARK 1.5. (i) The last part of the above proof is a particular instance of a general argument showing that certain Banach lattices are of cotype 2 (cf. [41], Theorem 3). Clearly (1.6) shows that B has a natural Banach lattice structure.

(ii) It is possible to see (cf. the proof of Corollary 1.3 in [42]) that Theorem 1.5 implies, by general arguments from Banach space theory, the main inequalities in Chapter 3.

We will give two consequences of the fact that B is of cotype 2.

COROLLARY 1.6. *Let* $\Gamma_1, \Gamma_2, \cdots, \Gamma_j, \cdots$ *be a partition of* Γ *into disjoint subsets, then we have for all* $a = \{a_\gamma | \gamma \epsilon \Gamma\} \epsilon B$

$$\sum_{j=1}^{\infty} E\|\sum_{\gamma\epsilon\Gamma_j} a_\gamma \varepsilon_\gamma \gamma\|^2 \leq (2C'C'')^2 E\|\sum_{\gamma\epsilon\Gamma} a_\gamma \varepsilon_\gamma \gamma\|^2 . \tag{1.10}$$

Proof. We define $a_\gamma^j = a_\gamma$ if $\gamma\epsilon\Gamma_j$ and $a_\gamma^j = 0$ otherwise. Also let $a^j = \{a_\gamma^j | \gamma\epsilon\Gamma\}$. Since $\left(\sum_{j=1}^{\infty} |a_\gamma^j|^2\right)^{1/2} = |a_\gamma|$, we have first by (1.7) (which clearly still holds for $N = \infty$) and then by (1.6) that

$$\begin{aligned}\sum_{j=1}^{\infty} |||a^j|||^2 &\leq (C'C'')^2 |||\{|a_\gamma|\}|||^2 \\ &\leq (2CC'')^2 |||a|||^2 .\end{aligned}$$

This inequality is equivalent to (1.10).

COROLLARY 1.7. *For any* $a = \{a_\gamma | \gamma\epsilon\Gamma\}$ *in* B, *there exist scalars* $\{\lambda_\gamma | \gamma\epsilon\Gamma\}$ *and* $\{\mu_\gamma | \gamma\epsilon\Gamma\}$ *such that*

$$a_\gamma = \lambda_\gamma \mu_\gamma \quad \textit{for all } \gamma\epsilon\Gamma$$

and verifying

(i) $\sum |\lambda_\gamma|^2 \leq 1$

and

(ii) *For all* $\{a_\gamma | \gamma \in \Gamma\} \in \ell_2(\Gamma)$, $\{a_\gamma \mu_\gamma | \gamma \in \Gamma\} \in B$ *and*

$$|||\{a_\gamma \mu_\gamma\}||| \leq C'C''\left(\sum |a_\gamma|^2\right)^{1/2}.$$

Proof. As usual, we denote $\ell_\infty(\Gamma)$ the Banach space of all bounded families $\{\beta_\gamma\}_{\gamma\in\Gamma}$ of scalars, and we denote $c_0(\Gamma)$ the closure in $\ell_\infty(\Gamma)$ of the set of finitely supported families $\{\beta_\gamma\}$. Fix $a = \{a_\gamma\}$ in B. By (2.4.9), we can define a bounded linear operator $T : c_0(\Gamma) \to B$ as follows:

$$T(\{\beta_\gamma\}) = \{\beta_\gamma a_\gamma | \gamma \in \Gamma\}.$$

It is not difficult to see, using (1.7), that T is 2-absolutely summing and $\pi_2(T) \leq C'C''|||a|||$, (for details see Theorem 3 in [41], and for more information on absolutely summing operators see e.g. [30]). Therefore, by Pietsch's factorization theorem (cf. Proposition 3.1 in [30]) we have, that there exists $\nu_\gamma \geq 0$ with $\sum_{\gamma\in\Gamma} \nu_\gamma \leq 1$ and such that, for all $\beta \in c_0(\Gamma)$,

$$\|T\beta\| \leq C'C''|||a|||\left(\sum_\Gamma \nu_\gamma |\beta_\gamma|^2\right)^{1/2}.$$

This is clearly equivalent to the announced result (take $\lambda_\gamma = (\nu_\gamma)^{1/2}$ and $\mu_\gamma = a_\gamma \lambda_\gamma^{-1}$ with the convention $0/0 = 0$).

CHAPTER V

RANDOM FOURIER SERIES ON COMPACT NON-ABELIAN GROUPS

1. *Introduction*

We consider random Fourier series on noncommutative compact groups, and extend practically all the results proved in Chapter III to this setting. The main result shows that the entropy condition is again a necessary and sufficient condition for the a.s. continuity of the generalized Rademacher or Steinhaus random Fourier series. The subject of random Fourier series in the non-Abelian case has been studied in the papers [11], [12], [13] and [52].

In this chapter, G will always be a compact group, not necessarily Abelian. Let m be the normalized Haar measure on G. We will denote by Σ the dual object to G, i.e. the set of all equivalence classes of irreducible representations on G. Since G is compact, all the representations in Σ are finite dimensional. For each i in Σ, we have a finite dimensional complex Hilbert space H_i, of dimension d_i, and an irreducible unitary representation $U_i : G \to B(H_i)$ which belongs to the class determined by i. By the classical Peter-Weyl Theorem (cf. [17] §27 and [57]), any function f in $L_2(G, m)$ (denoted simply $L_2(G)$) has a Fourier series expansion of the form

$$f(x) = \sum_{i \in \Sigma} d_i \operatorname{tr} U_i(x) \hat{f}(i)$$

where the sum is convergent in $L^2(G)$ with

$$\|f\|_2 = \left(\int |f|^2 dm \right)^{1/2} = \left[\sum_{i \in \Sigma} d_i \operatorname{tr} |\hat{f}(i)|^2 \right]^{1/2},$$

and where the Fourier "coefficient" $\hat{f}(i)$ of f is the element of $B(H_i)$ defined for all $i \in \Sigma$ by

$$\hat{f}(i) = \int U_i(-x) f(x) m(dx) .$$

In the sequel, whenever T is an operator on some Hilbert space, we denote by $|T|$ the operator $(T^*T)^{1/2}$. Let $U(H_i)$ be the group of all unitary operators on H_i. For future reference we select an orthonormal basis $\{e_j^i | 1 \le j \le d_i\}$ of H_i. We can now introduce the various types of random variables which generalize the Steinhaus, Rademacher and Gaussian (real or complex) random variables. All the random variables are defined on some probability space $(\Omega, \mathcal{F}, P)$. We denote by $\{W_i\}_{i \in \Sigma}$ a collection of independent random variables, each W_i taking values in $U(H_i)$ and being uniformly distributed on $U(H_i)$ (i.e. the distribution of $\omega \to W_i(\omega)$ coincides with the normalized Haar measure of $U(H_i)$). We also consider a collection $\{\varepsilon_i\}_{i \in \Sigma}$ of independent random variables, each ε_i being a random $d_i \times d_i$ orthogonal matrix (i.e. ε_i is a random matrix with *real* entries) uniformly distributed on the orthogonal group $O(d_i)$. The variables $\{\varepsilon_i\}_{i \in \Sigma}$ are the analogue in our new setting of Rademacher (or Bernoulli) random variables, while the variables $\{W_i\}_{i \in \Sigma}$ are the analogue of the Steinhaus random variables. As in the Abelian case, we will see that it is equivalent in our study to work with $\{\varepsilon_i\}$ or with $\{W_i\}$. Since we have selected (once and for all) a basis of H_i, we may as well consider ε_i as a random operator on H_i rather than as a random matrix. More generally, we will always identify in what follows a $d_i \times d_i$ matrix with the corresponding operator on H_i.

Let $\{g_{jk}^i | 1 \le j, k \le d_i, i \in \Sigma\}$ (resp. $\{\tilde{g}_{jk}^i | 1 \le j, k \le d_i, i \in \Sigma\}$) be a collection of independent real (resp. complex) valued Gaussian random variables with mean zero and variance 1. We denote G_i (resp. $\tilde{G}_i$) the random operator on H_i which admits $\left\{\frac{1}{\sqrt{d_i}} g_{jk}^i | 1 \le j, k \le d_i\right\}$ (resp. $\left\{\frac{1}{\sqrt{d_i}} \tilde{g}_{jk}^i\right\}$) as its representative matrix on the basis $\{e_j^i | 1 \le j \le d_i\}$, so that we have

$$G_i e_k^i = \frac{1}{\sqrt{d_i}} \sum_{j=1}^{d_i} g_{jk}^i e_j^i .$$

Note that $\{G_i\}_{i \epsilon \Sigma}$ (resp. $\{\tilde{G}_i\}_{i \epsilon \Sigma}$) is an independent family of random operators. For convenience we include some elementary facts on these variables:

PROPOSITION 1.1.

(i) *Let* $\{T_i\}_{i \epsilon \Sigma}$ *be a family of operators with* $T_i \epsilon B(H_i)$ *for each* i *in* Σ. *Assume that* $\Sigma\, d_i \mathrm{tr}\, T_i^* T_i < \infty$. *Then the series* $\sum_{i \epsilon \Sigma} d_i \mathrm{tr}\, T_i \varepsilon_i$, $\sum_{i \epsilon \Sigma} d_i \mathrm{tr}\, T_i W_i$, $\sum_{i \epsilon \Sigma} d_i \mathrm{tr}\, T_i G_i$ *and* $\sum_{i \epsilon \Sigma} d_i \mathrm{tr}\, T_i \tilde{G}_i$ *are all convergent in* $L^2(dP)$ *and the norms in* $L^2(dP)$ *of their sums are all equal to* $\left(\sum_{i \epsilon \Sigma} d_i \mathrm{tr}\, T_i^* T_i\right)^{1/2}$.

(ii) *Let* A_i. B_i *be both orthogonal (resp. both unitary)* $d_i \times d_i$ *matrices, considered as operators on* H_i; *then the distribution of the family of random variables* $\{A_i \varepsilon_i B_i\}_{i \epsilon \Sigma}$ *(resp.* $\{A_i W_i B_i\}_{i \epsilon \Sigma}$*) is the same as that of the original family* $\{\varepsilon_i\}_{i \epsilon \Sigma}$ *(resp.* $\{W_i\}_{i \epsilon \Sigma}$*).*

(iii) *Similarly,* $\{A_i G_i B_i\}_{i \epsilon \Sigma}$ *(resp.* $\{A_i \tilde{G}_i B_i\}_{i \epsilon \Sigma}$*) has the same distribution as* $\{G_i\}_{i \epsilon \Sigma}$ *(resp.* $\{\tilde{G}_i\}_{i \epsilon \Sigma}$*).*

Proof. Part i) follows from Parseval's equality. (We recall the fact that the random variables $\{g_{jk}^i\}_{i \epsilon \Sigma, 1 \leq j,k \leq d_i}$ are orthonormal in $L^2(dP)$ and similarly for the Steinhaus and Rademacher random variables.

Parts ii) and iii) are the analogue of the familiar sign invariance of Rademacher or Gaussian variables. The case of $\{\varepsilon_i\}_{i \epsilon \Sigma}$ and $\{W_i\}_{i \epsilon \Sigma}$ is obvious. We check (iii) for the case of e.g. $\{G_i\}_{i \epsilon \Sigma}$. Obviously, it is enough to show that G_i and $A_i G_i B_i$ have the same distribution for each fixed i in Σ, but then since both G_i and $A_i G_i B_i$ are (vector valued) Gaussian variables, it is enough to show that their covariances are the same. This follows immediately from Part i), since

$$E|\mathrm{tr}\ T_iA_iG_iB_i|^2 = E|\mathrm{tr}\ B_iT_iA_iG_i|^2 = (d_i)^{-1}\mathrm{tr}|B_iT_iA_i|^2$$
$$= (d_i)^{-1}\mathrm{tr}|T_i|^2 = E|\mathrm{tr}\ T_iG_i|^2 .$$

REMARK 1.2. By a similar argument, we can observe that $\{\varepsilon_i\}_{i\in\Sigma}$ has the same distribution as $\{\varepsilon_i^*\}_{i\in\Sigma}$, and similarly for $\{W_i\}_{i\in\Sigma}$, $\{G_i\}_{i\in\Sigma}$ and $\{\tilde{G}_i\}_{i\in\Sigma}$.

REMARK 1.3. It clearly follows from Proposition 1.1 and the polar decomposition theorem that if $T_i \in B(H_i)$ for each $i \in \Sigma$, then the distribution of $\{W_iT_i\}_{i\in\Sigma}$ is the same as that of $\{W_i|T_i|\}_{i\in\Sigma}$. A similar result applies to $\{\tilde{G}_i\}_{i\in\Sigma}$. Moreover, if the matrix associated with each T_i has real entries then a similar result also holds for $\{\varepsilon_i\}_{i\in\Sigma}$ and $\{G_i\}_{i\in\Sigma}$.

REMARK 1.4. Let $\{\tilde{G}_i^1\}_{i\in\Sigma}, \{\tilde{G}_i^2\}_{i\in\Sigma}, \cdots$ be a sequence of independent copies of the family $\{\tilde{G}_i\}_{i\in\Sigma}$. Then, by a well-known property of Gaussian variables, for each integer N, the family $\left\{\frac{1}{\sqrt{N}}\sum_{j=1}^{N} \tilde{G}_i^j \,|\, i\in\Sigma\right\}$ has the same distribution as that of $\{\tilde{G}_i\}_{i\in\Sigma}$. In fact, given $T_i^1, T_i^2, \cdots, T_i^N$, N operators in $B(H_i)$ for each $i\in\Sigma$, the family $\left\{\sum_{j=1}^{N} \tilde{G}_i^j T_i^j \,|\, i\in\Sigma\right\}$ has the same distribution as the family $\left\{\tilde{G}_i\left(\sum_{j=1}^{N} |T_i^j|^2\right)^{1/2} \,|\, i\in\Sigma\right\}$. A similar remark holds in the real case for $\{G_i\}$. This follows by the same argument as Proposition 1.1 (i.e. by comparing the covariances).

The following estimate plays an important role, although it is known, we do not have a suitable reference.

PROPOSITION 1.5. *Let* $\{g_{jk}\}_{1\le j,k\le d}$ *be an independent family of Gaussian real or complex random variables with mean zero and variance* 1.

Denote by G_d *the* $d \times d$ *matrix* $G_d = \left(\frac{1}{\sqrt{d}} g_{jk}\right)_{1 \le j,k \le d}$ *and denote* $\|G_d\|_\infty$, $\|G_d\|_2$, *and* $\|G_d\|_1$ *respectively the operator norm, the Hilbert Schmidt norm, and the nuclear norm (also called the trace class norm) of* G_d *considered as an operator acting on the* d*-dimensional (real or complex) Euclidean space denoted by* H. *These norms can be estimated as follows:*

(1.1) $$(E\|G_d\|_2^2)^{1/2} = \sqrt{d}$$

(1.2) $$\delta \le E\|G_d\|_\infty \le \alpha$$

(1.3) $$\delta d \le E\|G_d\|_1 \le d$$

where α *and* $\delta > 0$ *are absolute constants (independent of* d *).*

Proof. For convenience, we give the proof in the real case, the complex case is entirely similar. Let B denote the unit ball of H. As is well known, for each $\varepsilon > 0$, we can find an ε-net S of B with $|S| = \text{card}(S) \le (1+2/\varepsilon)^d$. Observe that

$$\|G_d\|_\infty = \sup\{<G_d x, y> x, y \in B\}.$$

Given x, y in B, there are x', y' in S with $\|x-x'\| \le \varepsilon$ and $\|y-y'\| \le \varepsilon$. Therefore

$$<G_d x, y> = <G_d x', y'> + <G_d(x-x'), y'> + <G_d x, y-y'>$$
$$\le \sup_{x', y' \in S} <G_d x', y'> + 2\varepsilon\|G_d\|_\infty .$$

This shows that

$$\|G_d\|_\infty \le \frac{1}{1-2\varepsilon} \sup_{x', y' \in S} <G_d x', y'> .$$

To complete the proof we use the following well-known lemma which is an immediate consequence of the Borel-Cantelli lemma.

LEMMA 1.6. *Let* $g_1, \cdots, g_N$ *be a collection of (not necessarily independent) Gaussian random variables. There is a universal constant* C *such that*

$$E \sup_{i \le N} |g_i| \le C(1+\operatorname{Log} N)^{1/2} \sup_{i \le N} (E|g_i|^2)^{1/2} .$$

We now end the proof of Proposition 1.5. We have

$$E\|G_d\|_\infty \le \frac{1}{1-2\varepsilon} E \sup_{x',y' \in S} \langle G_d x', y' \rangle$$

$$\le \frac{C}{1-2\varepsilon} (1+2 \operatorname{Log} |S|)^{1/2} \sup_{x',y' \in S} (E|\langle G_d x', y' \rangle|^2)^{1/2} .$$

Observe that, (denoting $\{x'_k\}_{k=1}^d$ and $\{y'_k\}_{k=1}^d$ the coordinates of x' and y'),

$$E|\langle G_d x', y' \rangle|^2 = \frac{1}{d} E\Big|\sum_{j,k} g_{jk} x'_k y'_j\Big|^2$$

$$= \frac{1}{d}\Big(\sum_k |x'_k|^2\Big)\Big(\sum_j |y'_j|^2\Big) \le \frac{1}{d} .$$

Therefore, taking for instance $\varepsilon = 1/4$, we have $|S| \le 9^d$ and we obtain $E\|G_d\|_\infty \le \alpha$, where α is an absolute constant. This establishes the right side of (1.2). The equality (1.1) is obvious since

$$E\|G_d\|_2^2 = E\Big(\frac{1}{d}\sum_{j,k} |g_{jk}|^2\Big) = d .$$

We turn to (1.3); recall that for any operator T we have,

$$\|T\|_2 \le (\|T\|_\infty \|T\|_1)^{1/2} .$$

Hence by the Cauchy-Schwarz inequality

$$E\|G_d\|_2 \le (E\|G_d\|_1)^{1/2} (E\|G_d\|_\infty)^{1/2} . \tag{1.4}$$

On the other hand, we have

$$E\|G_d\|_2 = \frac{1}{\sqrt{d}}\, E\Big(\sum_{j,k} |g_{jk}|^2\Big)^{1/2} \geq \frac{1}{\sqrt{d}}\Big\{\sum_{j,k} (E|g_{jk}|)^2\Big\}^{1/2} = \sqrt{d}\,\sqrt{2/\pi}\ .$$

Therefore, combining (1.4) and the right side of (1.2), we find

$$E\|G_d\|_1 \geq \Big(\frac{2}{\pi a}\Big) d\ .$$

Moreover, we have trivially $\|G_d\|_1 \leq d\|G_d\|_\infty$ so that $E\|G_d\|_\infty \geq \frac{2}{\pi a}$ and also $\|G_d\|_1 \leq \sqrt{d}\,\|G_d\|_2$ so that $(E\|G_d\|_1^2)^{1/2} \leq d$. This completes the proof of Proposition 1.5 (with $\delta = \frac{2}{\pi a}$).

REMARK 1.7. Let ξ be a real (resp. complex) random $d \times d$ matrix, considered as an operator on a d dimensional real (resp. complex) Hilbert space. Assume that for any orthogonal (resp. unitary) matrix U, $U\xi U^*$ has the same distribution as ξ. Then the matrix $E|\xi|$ has to be a scalar multiple of the identity. Indeed, we have

$$U(E|\xi|)U^* = E(U|\xi|U^*) = E(|U\xi U^*|) = E|\xi|\ .$$

Therefore, $E|\xi|$ commutes with any orthogonal (resp. unitary) matrix, so it must be a scalar multiple of the identity.

As an immediate consequence, we have

COROLLARY 1.8. *With the notation of Proposition 1.5, we have*

$$E|G_d| = \delta(d)\,\mathrm{Id}_H$$

where $\delta(d)$ *is a real number satisfying* $\delta(d) \geq \delta$ *for some absolute constant* $\delta > 0$ *independent of* d *and* Id_H *is the identity operator on* H.

Proof. By Remark 1.7 we know that $E|G_d| = \delta(d)\mathrm{Id}_H$ for some $\delta(d)$. Hence $\mathrm{tr}\, E|G_d| = \delta(d)d$, but by (1.3) we have

$$\mathrm{tr}\, E|G_d| = E\,\mathrm{tr}|G_d| = E\|G_d\|_1 \geq \delta d\ .$$

Therefore $\delta(d) \geq \delta > 0$.

2. *Random series with coefficients in a Banach space*

Before considering random Fourier series, it will be worthwhile to make a preliminary study of general random series in an arbitrary Banach space (real or complex) B. We will use the following convenient notation. Let H be a finite dimensional (real or complex) Hilbert space of dimension d and let $\{e_j | 1 \leq j \leq d\}$ be an orthonormal basis of H. Given a matrix $x = \{x_{jk} | 1 \leq j, k \leq d\}$ with entries in B, and an operator A in B(H) represented with respect to $\{e_j\}$ by the matrix $\{a_{jk} | 1 \leq j, k \leq d\}$. We will denote tr Ax, or tr xA, the element of B defined by

$$\operatorname{tr} xA = \operatorname{tr} Ax = \sum_{j,k=1}^{d} a_{jk} x_{kj} .$$

Also, we denote by $\|A\|_\infty$ the norm of A in B(H). (In a more sophisticated way, we could consider x and A as elements of $H \otimes B$ and $H \otimes H$, and use the natural pairing:

$$H \otimes H \times H \otimes B \to H \otimes B$$

to define this generalized notion of trace.)

In this section, Σ will be an arbitrary index set, $\{H_i\}_{i \in \Sigma}$ will be an arbitrary family of finite dimensional real or complex Hilbert spaces with dimensions $\{d_i\}_{i \in \Sigma}$. As before, we select an orthonormal basis in each H_i so that the notation introduced at the beginning of this section makes sense in each H_i. Moreover, the random variables ε_i, W_i, G_i, and $\tilde{G}_i$ can be defined as in the preceding section. For each i in Σ, we will consider a matrix $x^i = \{x^i_{jk} | 1 \leq j, k \leq d_i\}$ with entries in the Banach space B. Throughout this section F will be a fixed finite subset of Σ.

We will study the B-valued random variables

$$Y(\omega) = \sum_{i \in F} d_i \operatorname{tr} \varepsilon_i(\omega) x^i \tag{2.1}$$

and

$$\tilde{Y}(\omega) = \sum_{i \in F} d_i \operatorname{tr} W_i(\omega) x^i . \tag{2.2}$$

These are the analogues of the familiar Rademacher and Steinhaus series.

In the next proposition, we consider a family of random operators $\{V_i\}_{i\in\Sigma}$ with V_i taking its values in $B(H_i)$ for each i in Σ. We assume moreover that the distribution of $\{V_i\}_{i\in\Sigma}$ satisfies the following symmetry property, which is a generalization of the concept of sign-invariant random variables.

(2.3) For arbitrary a^i in $B(H_i)$ with a^i unitary ("unitary" means orthogonal in the real case), the family $\{a^iV_i\}_{i\in\Sigma}$ has the same distribution as $\{V_i\}_{i\in\Sigma}$.

We have the following version of the "contraction principle."

PROPOSITION 2.1. *Let* T^i *be an arbitrary operator in* $B(H_i)$, *for* $i\in\Sigma$. *Then the following inequalities hold, for each* $p\geq 1$:

$$\left(E\|\sum_{i\in F} \operatorname{tr} T^iV_ix^i\|^p\right)^{1/p} \leq \sup_{i\in\Sigma} \|T^i\|_\infty \left(E\|\sum_{i\in F} \operatorname{tr} V_ix^i\|^p\right)^{1/p}. \tag{2.4}$$

If T^i *is invertible for each* i, *then*:

$$\left(E\|\sum_{i\in F} \operatorname{tr} V_ix^i\|^p\right)^{1/p} \leq \sup_{i\in F} \|(T^i)^{-1}\|_\infty \left(E\|\sum_{F} \operatorname{tr} T^iV_ix^i\|^p\right)^{1/p}. \tag{2.5}$$

Moreover, we have in the real case (i.e. when H_i *is a real Hilbert space)*

$$E\|\sum_{i\in F} \operatorname{tr} \varepsilon_i E|V_i|x^i\|^p \leq E\|\sum_{i\in F} \operatorname{tr} V_ix^i\|^p \tag{2.6}$$

and

$$\left(E\|\sum_{i\in F} \operatorname{tr} \varepsilon_i x^i\|^p\right)^{1/p} \leq \sup_{F} \|(E|V_i|)^{-1}\|_\infty \left(E\|\sum_{F} \operatorname{tr} V_ix^i\|^p\right)^{1/p}. \tag{2.7}$$

In the complex case, the last two inequalities hold with $\{W_i\}$ *in place of* $\{\varepsilon_i\}$.

Proof. We give the proof in the real case only, the complex case is entirely similar. Let K_i be the unit ball of $B(H_i)$ and let $K = \prod_{i\in F} K_i$.

By homogeneity it is enough to prove (2.4) for $\{T^i\}_{i\epsilon F}$ in K. Clearly K is a compact convex set and, as is well known, its extreme points are exactly of the form $\{a^i\}_{i\epsilon F}$ with a^i unitary for every i in F. Consider the function $f: K \to \mathbb{R}$ defined by $f(\{T^i\}) = (E\|\Sigma \operatorname{tr} T^i V_i x^i\|^p)^{1/p}$. Since f is convex, the Krein-Milman theorem shows that f attains its supremum on an extreme point of K; but if $\{a^i\}_{i\epsilon F}$ is such an extreme point, then (2.3) implies that

$$f(\{a^i\}) = \left(E\|\sum \operatorname{tr} V_i x^i\|^p\right)^{1/p}.$$

Therefore, we must have

$$\sup\{f(x) | x \epsilon K\} \leq \left(E\|\sum \operatorname{tr} V_i x^i\|^p\right)^{1/p}$$

which is the desired inequality (2.4). The next inequality (2.5) follows immediately from (2.4). We turn to (2.6). We assume without restricting the generality that $\{\varepsilon_i\}_{i\epsilon\Sigma}$ is independent of $\{V_i\}_{i\epsilon\Sigma}$. We need the observation that the distribution of $\{V_i\}_{i\epsilon\Sigma}$ is the same as that of $\{\varepsilon_i|V_i|\}_{i\epsilon\Sigma}$. The reader can check this fact using Remark 1.3. Now, denote E_ε and E_V respectively the expectation signs with respect to $\{\varepsilon_i\}$ and $\{V_i\}$. We have

$$\begin{aligned} E\|\sum \operatorname{tr} V_i x^i\|^p &= E_\varepsilon E_V \|\sum \operatorname{tr} \varepsilon_i |V_i| x^i\|^p \\ &\geq E_\varepsilon\|\sum \operatorname{tr} \varepsilon_i E|V_i| x^i\|^p \end{aligned}$$

which establishes (2.6).

Finally, (2.7) is a consequence of (2.6) and the following claim:

$$E\|\sum \operatorname{tr} \varepsilon_i x^i\|^p \leq (\sup\|(E|V_i|)^{-1}\|_\infty)^p E\|\sum \operatorname{tr} \varepsilon_i E|V_i| x^i\|^p .$$

This claim follows from (2.5) applied to the case $V_i = \varepsilon_i$ provided we take into account Remark 2.2 below.

REMARK 2.2. Clearly, Proposition 2.1 remains true with $V_i T_i$ instead of $T_i V_i$ as long as a similar modification is made in the hypothesis (2.3).

REMARK 2.3. If $\{\varepsilon_i\}_{i\epsilon\Sigma}$ is independent of $\{V_i\}_{i\epsilon\Sigma}$, then $\{V_i\}_{i\epsilon\Sigma}$ verifies the hypothesis (2.3) if and only if $\{V_i\}_{i\epsilon\Sigma}$ has the same distribution as $\{\varepsilon_i V_i\}_{i\epsilon\Sigma}$ ($\{W_i V_i\}_{i\epsilon\Sigma}$ in the complex case).

The next corollary is quite useful, it follows immediately from (2.7) and Corollary 1.8.

COROLLARY 2.4. *We have, for* $1 \le p < \infty$,

$$\delta^p E\|\sum_{i\in F} d_i \operatorname{tr} \varepsilon_i x^i\|^p \le E\|\sum_{i\in F} d_i \operatorname{tr} G_i x^i\|^p \tag{2.8}$$

where $\delta > 0$ *is an absolute constant. A similar result holds in the complex case with* $\{W_i\}$ *and* $\{\tilde{G}_i\}$ *in the place of* $\{\varepsilon_i\}$ *and* $\{G_i\}$.

REMARK 2.5. In the complex case, although the operators $\{T^i\}$ have a matrix with *complex* entries, we still have

$$\left(E\|\sum_F \operatorname{tr} T^i \varepsilon_i x^i\|^p\right)^{1/p} \le 2\left(E\|\sum_F \operatorname{tr} \varepsilon_i x^i\|^p\right)^{1/p} \sup_{i\in F} \|T^i\|_\infty . \tag{2.9}$$

Indeed, we may write $T^i = a^i + \sqrt{-1}\, b^i$ where a^i, b^i have associated matrices with *real* entries and $\|a^i\|_\infty$ and $\|b^i\|_\infty$ are less than $\|T^i\|_\infty$. Then inequality (2.9) follows from Proposition 2.1 and the triangle inequality.

The next proposition shows that $\{\varepsilon_i\}_{i\in\Sigma}$ and $\{W_i\}_{i\in\Sigma}$ are essentially equivalent for our study.

PROPOSITION 2.6. *The following inequality holds if* $1 \le p < \infty$:

$$\frac{1}{2}\left(E\|\sum_{i\in F} \operatorname{tr} \varepsilon_i x^i\|^p\right)^{1/p} \le \left(E\|\sum_{i\in F} \operatorname{tr} W_i x^i\|^p\right)^{1/p} \le 2\left(E\|\sum_F \operatorname{tr} \varepsilon_i x^i\|^p\right)^{1/p} . \tag{2.10}$$

Proof. We may of course assume that $\{\varepsilon_i\}_{i\in\Sigma}$ and $\{W_i\}_{i\in\Sigma}$ are independent. We use the fact that $\{W_i\varepsilon_i\}_{i\in\Sigma}$ has the same distribution as $\{W_i\}_{i\in\Sigma}$ so that

$$E\|\sum_F \operatorname{tr} W_i \varepsilon_i x^i\|^p = E\|\sum_F \operatorname{tr} W_i x^i\|^p .$$

Since $\{\varepsilon_i\}$ and $\{W_i\}$ are independent, this implies that there exists $\omega_0 \in \Omega$ such that

$$E\|\sum_F \operatorname{tr} W_i x^i\|^p \leq E\|\sum_F \operatorname{tr} W_i(\omega_0)\varepsilon_i x^i\|^p .$$

By (2.9), this is less than $2^p\, E\|\sum_F \operatorname{tr} \varepsilon_i x^i\|^p$. This yields the right side of (2.10).

To prove the left side, we first find $\omega_0 \in \Omega$ such that

$$E\|\sum_F \operatorname{tr} W_i x^i\|^p \geq E\|\sum_F \operatorname{tr} W_i(\omega_0)\varepsilon_i x^i\|^p .$$

Then we write

$$E\|\sum_F \operatorname{tr} \varepsilon_i x^i\|^p = E\|\sum_F \operatorname{tr} W_i(\omega_0)W_i(\omega_0)^* \varepsilon_i x^i\|^p$$

and again Remark 2.5 shows that this is less than $2^p E\|\sum_F \operatorname{tr} W_i(\omega_0)\varepsilon_i x^i\|^p$, which yields the left side of (2.10).

The main result of this section is the next theorem (in the scalar case, this is known, cf. [17] §36.2). Since the dimensions d_i may be unbounded, the result does not seem to follow from the general results (cf. e.g. [18]) on the integrability of the norms of sums of independent vector-valued random variables.

THEOREM 2.7. *There are absolute constants* $\varepsilon > 0$ *and* K *such that any variable* Y *, of the form (2.1), which verifies*

$$P(\{\|Y\| > 1\}) \leq \varepsilon$$

must also verify

$$E \exp\left(\frac{\|Y\|^2}{K^2}\right) \leq 2 .$$

A similar statement holds with $\tilde{Y}$ *instead of* Y .

In the particular case $d_i = 1$ for all i, this result is due to Kwapień [28], who improved a previous result of Kahane [26]. Kwapień's proof depends on Kahane's result and an additional argument. In the proof below, we first prove a suitable generalization of the latter argument and then we show how this argument alone is enough to complete the proof.

The first lemma is elementary.

LEMMA 2.8. *Let* f *be a positive random variable. Define for* $\alpha > 0$

$$N_\alpha(f) = \inf\{c > 0 \mid E \exp(f/c)^\alpha \le 2\} .$$

Then, if $P(\{f > 1\}) \le \varepsilon < 1/2$, *we have*

$$N_1(f) \le 2 + \phi(\varepsilon) N_2(f)$$

where $\phi(\varepsilon) = \left(\mathrm{Log}\, \frac{1}{2\varepsilon}\right)^{-1/2}$ *so that, in particular,* $\phi(\varepsilon) \to 0$ *when* $\varepsilon \to 0$.

Proof. We first write

$$\text{(2.10')} \qquad N_1(f) \le N_1(f\, 1_{\{f \le 1\}}) + N_1(f\, 1_{\{f > 1\}})$$
$$\le 2 + N_1(f\, 1_{\{f>1\}}) .$$

We claim that if $P(\{f > 1\}) \le \varepsilon$ then

$$N_1(f\, 1_{\{f>1\}}) \le \phi(\varepsilon) N_2(f) .$$

To prove this claim, let $a = N_2(f)$, we have for all $c > 0$

$$E \exp\left(\frac{f}{ca} 1_{\{f>1\}}\right) \le 1 + \int_{\{f>1\}} \exp\left(\frac{f}{ca}\right) dP$$

$$\le 1 + \sqrt{\varepsilon}\left(\int \exp\left(\frac{2f}{ca}\right) dP\right)^{1/2} .$$

(The last inequality follows from the Cauchy-Schwarz inequality.) We have, obviously, $\exp\left(\frac{2f}{ca}\right) \le \exp\left(\frac{1}{c^2} + \frac{f^2}{a^2}\right)$ so that

$$\mathrm{E}\exp\left(\frac{f}{ca}1_{\{f>1\}}\right) \le 1 + \sqrt{\varepsilon}\left(\exp\frac{1}{c^2}\,\mathrm{E}\exp\frac{f^2}{a^2}\right)^{1/2},$$

and since $a = N_2(f)$ this is

$$\le 1 + \left(\sqrt{2\varepsilon}\exp\left(\frac{1}{2c^2}\right)\right).$$

Finally, choosing $c = \phi(\varepsilon)$, we have

$$\mathrm{E}\exp\left(\frac{f}{ca}1_{\{f>1\}}\right) \le 2$$

so that $N_1(f\,1_{\{f>1\}}) \le \phi(\varepsilon)a = \phi(\varepsilon)N_2(f)$. Together with (2.10′) this proves the lemma.

In the next lemma, we consider a sequence $\{\varepsilon_i^1\}_{i\epsilon\Sigma}, \{\varepsilon_i^2\}_{i\epsilon\Sigma}, \cdots, \{\varepsilon_i^N\}_{i\epsilon\Sigma}, \cdots$ of independent copies of $\{\varepsilon_i\}_{i\epsilon\Sigma}$ (and similarly for $\{W_i\}_{i\epsilon\Sigma}$). We then have

LEMMA 2.9. *There is an absolute constant* $\delta_1 > 0$ *such that*

$$\mathrm{E}\,\mathrm{tr}\left|\sum_{k=1}^{N}\varepsilon_i^k\right| \ge \delta_1 N^{1/2} d_i$$

for any i *in* Σ *and any integer* N. *A similar result holds for* $\{W_i\}$.

Proof. The above inequality follows easily from Proposition 3.2 in [56]; we provide a similar argument for the sake of completeness. For simplicity, we fix i and N and write simply $S = \sum_{k=1}^{N}\varepsilon_i^k$, and

$$\delta_k' = \frac{1}{2}(\varepsilon_i^k + \varepsilon_i^{k*});\quad \delta_k'' = \frac{1}{2\sqrt{-1}}(\varepsilon_i^k - \varepsilon_i^{k*}).$$

It will be convenient to work here with a complex Hilbert space H_i, so that we can define the random operators π', π'' and π on H_i as follows:

$$\pi' = \prod_{k=1}^{N} (N^{-\frac{1}{2}}\delta'_k + \sqrt{-1}\ \mathrm{Id}_{H_i}) ,$$

$$\pi'' = \prod_{k=1}^{N} (N^{-\frac{1}{2}}\delta''_k + \sqrt{-1}\ \mathrm{Id}_{H_i}) ,$$

$$\pi = \pi' + \sqrt{-1}\ \pi'' .$$

We have, clearly,

$$\|N^{-\frac{1}{2}}\delta'_k + \sqrt{-1}\ \mathrm{Id}_{H_i}\|_\infty \le (1+N^{-1}\|\delta'_k\|_\infty^2)^{\frac{1}{2}} \le (1+N^{-1})^{\frac{1}{2}}$$

so that

$$\|\pi'\|_\infty \le \prod_{k=1}^{N} \|N^{-\frac{1}{2}}\delta'_k + \sqrt{-1}\ \mathrm{Id}_{H_i}\|_\infty$$

$$\le (1+N^{-1})^{N/2} \le \sqrt{e} .$$

Similarly $\|\pi''\|_\infty \le \sqrt{e}$, so that $\|\pi\|_\infty \le 2\sqrt{e}$. On the other hand, since $\{\varepsilon_i^k\}_k$ are independent variables with mean zero, it is easy to check that $S\pi^* = N^{\frac{1}{2}}\mathrm{Id}_{H_i} + R$, where R is a certain sum of random operators with mean zero. Therefore, we have,

$$\begin{aligned} N^{\frac{1}{2}}d_i = E\ \mathrm{tr}\ S\pi^* &\le E\ \mathrm{tr}\ |S\pi^*| \\ &\le E\ \mathrm{tr}\ |S|\ \|\pi\|_\infty \\ &\le 2\sqrt{e}\ E\ \mathrm{tr}\ |S| , \end{aligned}$$

and this yields Lemma 2.9 with $\delta_1 = (2\sqrt{e})^{-1}$. The case of $\{W_i\}$ is entirely similar.

Now we can prove the main step.

LEMMA 2.10. *Let* Y *be as in (2.1). Then there is a universal constant* K_1 *such that*

$$N_2(\|Y\|) \le K_1 N_1(\|Y\|) .$$

A similar result holds for $\tilde{Y}$ *instead of* Y.

Proof. Following [28] we consider

$$Y^k = \sum_{i\in F} d_i \operatorname{tr} \varepsilon_i^k x^i \quad \text{for} \quad k = 1, 2, \cdots ,$$

where the variables $Y^1, Y^2, \cdots$ are independent copies of Y. Denote by S^N the sum $S^N = \sum_{k=1}^{N} Y^k$. Applying (2.6) we obtain, for all $p \geq 1$,

$$\text{(2.11)} \qquad E\|\sum_F d_i \operatorname{tr} \varepsilon_i E|\sum_{k=1}^{N} \varepsilon_i^k| x^i\|^p \leq E\|S^N\|^p .$$

Let us recall that by Remark 1.7, $E|\sum_{k=1}^{N} \varepsilon_i^k|$ is a multiple of the identity, say $E|\sum_{k=1}^{N} \varepsilon_i^k| = \lambda(i,N)\, Id_{H_i}$. If we take the trace of both sides of this equality and use Lemma 2.9 we get

$$\lambda(i,N) d_i = E \operatorname{tr} |\sum_{k=1}^{N} \varepsilon_i^k| \geq \delta_1 N^{1/2} d_i$$

so that $\lambda(i,N) \geq \delta_1 N^{1/2}$. Returning to (2.11) we have

$$E\|\sum_F d_i \operatorname{tr} \varepsilon_i x^i \lambda(i,N)\|^p \leq E\|S^N\|^p$$

and by the usual contraction principle (2.4.7) (a particular case of (2.7)) we have

$$\text{(2.12)} \qquad E\|Y\|^p = E\|\sum d_i \operatorname{tr} \varepsilon_i x^i\|^p \leq (\inf_i |\lambda(i,N)|)^{-p} E\|S^N\|^p$$

$$\leq \delta_1^{-p} E\|S^N\|^p N^{-p/2} .$$

Since $x^p \leq p^p e^{-p} e^x$ for all $x \geq 0$, we have for any $\lambda \geq 0$

$$\|\frac{S^N}{\lambda}\|^p \leq p^p e^{-p} \exp\left[\frac{\|S^N\|}{\lambda}\right] .$$

Hence

$$E\|S^N\|^p \le \lambda^p p^p e^{-p} E \exp\left[\frac{\|S^N\|}{\lambda}\right]$$

$$\le (\lambda p e^{-1})^p E \exp\left(\lambda^{-1} \sum_1^N \|Y^k\|\right)$$

$$\le (\lambda p e^{-1})^p (E \exp \lambda^{-1} \|Y\|)^N .$$

Choosing $\lambda = N_1(\|Y\|)$, we obtain

$$N^{-p/2} E\|S^N\|^p \le (\lambda p e^{-1}/\sqrt{N})^p 2^N$$

and by (2.12)

$$(E\|Y\|^p)^{1/p} \le \left(\frac{p}{\sqrt{N}}\right) \frac{N_1(\|Y\|)}{e\delta_1} 2^{N/p} .$$

Hence, if we choose $p = N$, we conclude that for all $N \ge 1$

$$(E\|Y\|^N)^{1/N} \le 2(e\delta_1)^{-1} \sqrt{N}\, N_1(\|Y\|) . \tag{2.13}$$

Now the conclusion follows from the following elementary and well-known fact:

LEMMA 2.11. *For any positive random variable* f *we have*

$$N_2(f) \le C \sup_{n \ge 1} \frac{1}{\sqrt{n}} (E f^n)^{1/n} ,$$

where C *is an absolute constant. (To check this, write the expansion* $\exp t^2 = \sum_{n=0}^{\infty} \frac{t^{2n}}{n!}$ *and apply Stirling's formula.)*

The preceding lemma and (2.13) imply

$$N_2(\|Y\|) \le 2C(e\delta_1)^{-1} N_1(\|Y\|)$$

which concludes the proof of Lemma 2.10.

Proof of Theorem 2.7. Assume that $P(\{\|Y\| > 1\}) < \varepsilon < 1/2$. By Lemmas 2.10 and 2.8 we have

$$\begin{aligned} N_2(\|Y\|) &\le K_1 N_1(\|Y\|) \\ &\le 2K_1 + K_1\phi(\varepsilon) N_2(\|Y\|) . \end{aligned}$$

Therefore, if $\varepsilon > 0$ has been chosen small enough so that $K_1\phi(\varepsilon) \le \frac{1}{2}$ we have

$$N_2(\|Y\|) \le 2K_1 + \frac{1}{2} N_2(\|Y\|)$$

so that $N_2(\|Y\|) \le 4K_1$. This completes the proof of the theorem with $K = 4K_1$ and $\varepsilon < 1/2$ chosen so that $K_1\phi(\varepsilon) < \frac{1}{2}$. The proof with $\tilde{Y}$ in the place of Y is completely similar.

COROLLARY 2.12. *If* $0 < p < q < \infty$, *there is an absolute constant* K_{pq} *such that any random variable* Y *as in (2.1) satisfies*

$$(E\|Y\|^q)^{1/q} \le K_{pq}(E\|Y\|^p)^{1/p} . \tag{2.14}$$

Moreover $K_{pq} \in O(\sqrt{q})$ *when* $q \to \infty$ *for each fixed* p. *A similar result holds for* $\tilde{Y}$.

Proof. Clearly, it is enough to prove this when q is an even integer; moreover, by homogeneity it is enough to prove (2.14) assuming $E\|Y\|^p = 1$. We then have $P(\{\varepsilon^{1/p}\|Y\| > 1\}) \le \varepsilon$. Therefore, applying Theorem 2.7 to $\varepsilon^{1/p}Y$, we have

$$E \exp(\varepsilon^{2/p}\|Y\|^2 K^{-2}) \le 2$$

and a fortiori for all $k = 1, 2, \cdots$

$$E(\varepsilon^{1/p}\|Y\|K^{-1})^{2k} \le k! E \exp(\varepsilon^{2/p}\|Y\|^2K^{-2}) \le 2k!$$

so that

$$(E\|Y\|^{2k})^{1/2k} \le (2k!)^{1/2k} K\varepsilon^{-1/p} \le K\varepsilon^{-1/p}\sqrt{2k} .$$

This is the announced result for q an even integer.

REMARK 2.13. Denote simply $L_0(B)$ (resp. $L_p(B)$) the vector space of all B-valued random variables on $(\Omega, \mathcal{F}, P)$ (resp. with p-integrable norm). Let E be the subspace of $L^0(B)$ spanned by all the variables Y of the form (2.1) with $\{x^i_{jk}\}$ arbitrary elements of B. The above theorem can be rephrased as follows: The topology of convergence in probability (i.e. the natural topology of the space $L_0(B)$) induces on E the same topology as $L_p(B)$, for any $p > 0$. Moreover, these topologies are all equivalent to the topology associated with the norm

$$N_2(\|Y\|) = \inf\{c > 0 | E \exp\left(\frac{\|Y\|}{c}\right)^2 \leq 2\} .$$

Of course a similar result holds for the subspace $\tilde{E}$ spanned by all the variables $\tilde{Y}$ of the form (2.2).

The last theorem summarizes several results of this section:

THEOREM 2.14. *Let* $\{x^i_{jk} | i \epsilon \Sigma, 1 \leq j, k \leq d_i\}$ *be elements of* B. *Let* F_n *be an increasing sequence of finite subsets of* Σ *with* $\bigcup_n F_n = \Sigma$ *(so we assume here that* Σ *is countable). Set*

$$Z_n = \sum_{i \epsilon F_n} d_i \operatorname{tr} \varepsilon_i x^i$$

and

$$\tilde{Z}_n = \sum_{i \epsilon F_n} d_i \operatorname{tr} W_i x^i .$$

Then the following eight assertions are equivalent:

(i) *(resp. (i)′) The sequence* $\{Z_n\}$ *(resp.* $\{\tilde{Z}_n\}$ *) is a.s. convergent in* B.

(ii) *(resp. (ii)′) The sequence* $\{Z_n\}$ *(resp.* $\{\tilde{Z}_n\}$ *) converges in probability, i.e. in* $L_0(B)$.

(iii) *(resp. (iii)′) For some* p, $0 < p < \infty$, $\{Z_n\}$ *(resp.* $\{\tilde{Z}_n\}$ *) is convergent in* $L_p(B)$.

(iv) *(resp. (iv)′) For any* p, $0 < p < \infty$, $\{Z_n\}$ *(resp.* $\{\tilde{Z}_n\}$ *) is convergent in* $L_p(B)$.

Proof. The previous remark explains why (ii) $\Longleftrightarrow$ (iii) $\Longleftrightarrow$ (iv) and also why (ii)′ $\Longleftrightarrow$ (iii)′ $\Longleftrightarrow$ (iv)′. The equivalences (i) $\Longleftrightarrow$ (ii) and (i)′ $\Longleftrightarrow$ (ii)′ follow from (2.4.2); they are valid in general for sums of independent symmetric Banach space valued random variables (cf. [18] Theorem 2.4). To complete the proof, it is enough to show for instance that $\{Z_n\}$ converges in $L_1(B)$ if and only if $\{\tilde{Z}_n\}$ converges in $L_1(B)$. But this last assertion is an obvious corollary to Proposition 2.6.

REMARK 2.15. The preceding proof shows that Proposition 2.6 remains true for $0 < p < 1$ provided we change the constants in the corresponding inequality.

REMARK 2.16. Of course, the preceding theorem has an analogue with $\{G_i\}$ and $\{\tilde{G}_i\}$ in the place of $\{\varepsilon_i\}$ and $\{W_i\}$, but this follows from general results on Gaussian variables. The equivalence of $\{G_i\}$ and $\{\tilde{G}_i\}$ analogous to Proposition 2.6 is obvious (i.e. the proof of Lemma 2.4.12 extends immediately).

REMARK 2.17. With the notation of Theorem 2.14, we have:

(i) If $Z_n \to Z$ a.s. in norm then necessarily $E \exp \delta\|Z\|^2 < \infty$ for some $\delta > 0$. Indeed, this follows immediately from Theorem 2.7. In fact, Theorem 2.7 remains clearly true with Z in the place of the finite sum Y.

(ii) If $M = \sup_n \|Z_n\| < \infty$ a.s., then necessarily $E \exp \delta M^2 < \infty$ for some $\delta > 0$. This also follows from Theorem 2.7 combined with P. Levy's inequality (2.4.2). These two statements are still valid with $\{W_i\}_{i\in\Sigma}$ in the place of $\{\varepsilon_i\}_{i\in\Sigma}$.

3. *Continuity of random Fourier series*

Now we turn to the special case of random Fourier series. We use the notation of Section 1. In complete analogy with the Abelian case, we will consider several kinds of series: Real or complex Gaussian series

(3.1) $$X(t) = \sum_{i \in \Sigma} d_i \operatorname{tr} G_i U_i(t) \hat{f}(i)$$

and

(3.2) $$\tilde{X}(t) = \sum_{i \in \Sigma} d_i \operatorname{tr} \tilde{G}_i U_i(t) \hat{f}(i) .$$

generalized Rademacher series

(3.3) $$Y(t) = \sum_{i \in \Sigma} d_i \operatorname{tr} \varepsilon_i U_i(t) \hat{f}(i)$$

and generalized Steinhaus series

(3.4) $$\tilde{Y}(t) = \sum_{i \in \Sigma} d_i \operatorname{tr} W_i U_i(t) \hat{f}(i) .$$

Moreover, we will also consider a family $\{\xi_i\}_{i \in \Sigma}$ of random variables with $\xi_i(\omega) \in B(H_i)$ for each i in Σ. We *do not* assume that $\{\xi_i\}_{i \in \Sigma}$ is an independent family, but merely that $\{\xi_i\}_{i \in \Sigma}$ is independent of $\{\varepsilon_i\}_{i \in \Sigma}$ and of $\{G_i\}_{i \in \Sigma}$. We will then consider the random series

(3.5) $$Z(t) = \sum_{i \in \Sigma} d_i \operatorname{tr} \varepsilon_i \xi_i U_i(t) \hat{f}(i) .$$

Since the set of those i for which $\hat{f}(i) \neq 0$ is countable, it does not restrict the generality to assume that Σ is countable, so that G is metrizable (for details on this, see Corollary 28.11 in [17]). From now on, we do assume G metrizable and Σ countable. We start this section by some general observations:

REMARK 3.1. It is clearly equivalent to say that the random Fourier series (3.1) represents a.s. a continuous function or to say that the process $\{X(t)\}_{t \in G}$ has a version with continuous sample paths; and similarly for the random series (3.2), (3.3), (3.4) and (3.5). When this situation occurs, we will say briefly that the random series or the corresponding process is a.s. continuous, and we will not distinguish between $X(t)$, $Y(t)$, etc. and their continuous versions.

REMARK 3.2. The five processes defined above can all be viewed as sums of a sign-invariant sequence of C(G)-valued random variables. Therefore the generalized Ito-Nisio theorem (see Remark 2.4.4) insures that if one of these processes, say for example $Y(t)$, is a.s. continuous then necessarily the following holds:

For any increasing sequence A_n of finite subsets of Σ with $\bigcup_n A_n = \Sigma$, the sequence of the "partial sums"

$$Y^n(t) = \sum_{i \in A_n} d_i \operatorname{tr} \varepsilon_i U_i(t) \hat{f}(i) \tag{3.6}$$

converges a.s. in C(G).

REMARK 3.3. The results of the preceding section (especially Theorem 2.14) apply to the above random series with $B = C(G)$ and $x^i = U_i \hat{f}(i)$ considered as a matrix of elements of C(G). (Recall that the entries of the matrix $U_i(t)$ are continuous functions of t.) In particular, Theorem 2.14 and the preceding remark show that the series (3.3) is a.s. continuous if and only if (3.4) is also a.s. continuous. Moreover, the a.s. continuity of (for example) (3.3) is equivalent to the convergence of the "partial sums" (3.6) in $L_p(dP; C(G))$ for any $p < \infty$. Similarly (cf. Remark 2.16) the a.s. continuity of (3.1) is equivalent to that of (3.2) and this is equivalent to the convergence of the partial sums in $L_p(dP; C(G))$ for any $p < \infty$, by the Landau-Shepp-Fernique theorem (see Theorem 2.4.7).

To each f in $L_2(G)$, we may associate a pseudo-metric d defined for all $s, t \in G$ by

$$\begin{aligned} d(s,t) &= (E|X(s) - X(t)|^2)^{1/2} \\ &= \Big(\sum_{i \in \Sigma} d_i \operatorname{tr} |(U_i(s) - U_i(t))\hat{f}(i)|^2\Big)^{1/2} \\ &= \|f_s - f_t\|_2 \end{aligned}$$

where f_t is defined by $f_t(x) = f(x+t)$.

The important property of d is that it is left translation invariant, i.e. for all $s, t, x \in G$

$$d(x+t, x+s) = d(t, s) .$$

This implies (as in the Abelian case) that $m(\{x \in G | d(x,t) < \varepsilon\})$ is independent of the point t in G. As before, we denote $N(\varepsilon)$ the smallest number of open balls of radius ε for the pseudo-metric d which suffice to cover G.

The Dudley-Fernique theorem extends immediately in this setting.

THEOREM 3.4. *The entropy condition*

$$\int_0^\infty (\log N(\varepsilon))^{1/2} d\varepsilon < \infty \tag{3.7}$$

is necessary and sufficient for the a.s. continuity of the random Fourier series (3.1) or (3.2). Moreover, there are absolute constants $C_1, C_2 > 0$ *such that*

$$\frac{1}{C_2} \underline{I}(f) \le E\|X\|_{C(G)} \le C_1 \underline{I}(f) \tag{3.8}$$

with $\underline{I}(f)$ *defined by*

$$\underline{I}(f) = \|f\|_2 + \int_0^\infty (\log N(\varepsilon))^{1/2} ,$$

(and similarly for $\tilde{X}$ *).*

In the sequel, we will always denote simply $\|\cdot\|$ the norm in the space $C(G)$.

The modifications needed to pass from the Abelian case (discussed above in Chapter III) to the non-Abelian case are really minor. Notice however that the a.s. continuity of

$$X(t) = \sum d_i \operatorname{tr} G_i U_i(t) \hat{f}(i)$$

is a priori *not* equivalent to the a.s. continuity of the process

$$W(t) = \sum d_i \operatorname{tr} U_i(t) G_i \hat{f}(i) .$$

Actually, these processes have different associated metrics and it turns out that one of them may be a.s. continuous while the other is not (see Chapter VII, Section 3 for an example). Of course, in the special case when $\hat{f}(i) = \hat{f}(i)^*$ for each i, then they are a.s. continuous at the same time, since, in that case, $\{W(t)\}_{t \in G}$ has the same distribution as $\{\overline{X(-t)}\}_{t \in G}$.

Our main result is the following:

THEOREM 3.5:

(i) *The generalized Rademacher series (3.3) is a.s. continuous if and only if the corresponding Gaussian series (3.1) is a.s. continuous. Moreover, there are absolute constants* $C_1', C_2' > 0$ *such that*

$$\text{(3.9)} \qquad \frac{1}{C_1'} E\|X\| \leq E\|Y\| \leq C_2' E\|X\| .$$

(ii) *The entropy condition (3.7) is also a necessary and sufficient condition for the a.s. continuity of* $\{Y(t)\}_{t \in G}$ *and* $\underline{I}(f)$ *is equivalent to* $E\|Y\|$.

(iii) *More generally, if*

$$\lambda = \sup_{i \in \Sigma} \| E|\xi_i|^2 \|_\infty^{1/2} < \infty$$

the a.s. continuity of $\{Y(t)\}$ *implies the a.s. continuity of* $\{Z(t)\}$, *and we have*

$$\text{(3.10)} \qquad (E\|Z\|^2)^{1/2} \leq K \lambda E\|Y\|$$

for some absolute constant K. *Furthermore, if* $E|\xi_i|$ *is invertible in* $B(H_i)$ *for each* i *in* Σ *and if*

$$\mu = \sup_{i \in \Sigma} \| (E|\xi_i|)^{-1} \|_\infty < \infty$$

then $\{Y_t\}$ *is a.s. continuous if and only if* $\{Z_t\}$ *is a.s. continuous and we have*

$$(3.11)\qquad E\|Y\| \leq K'\mu E\|Z\|$$

for some absolute constant K′.

The proof of Theorem 3.5 follows closely the pattern of the proof in the Abelian case. Several lemmas are required to adapt the arguments of Chapter III.

LEMMA 3.6. *There is an absolute constant* C_3 *such that, for any finite subset* A *of* Σ,

$$(3.12)\qquad E\|\sum_{i\epsilon A} d_i \mathrm{tr}\, \varepsilon_i \xi_i U_i \hat{f}(i)\|^2 \leq (C_3)^2 E\|\sum_{i\epsilon A} d_i \mathrm{tr}\, G_i U_i \hat{f}(i)\|^2 \cdot \sup_{i\epsilon A} \|E|\xi_i|^2\| \,.$$

Proof. First recall that $\{\varepsilon_i\}$ and $\{\xi_i\}$ are assumed independent of each other. Define, for fixed $\omega \epsilon \Omega$ and $t \epsilon G$,

$$\sigma_\omega(t) = \Big(\sum_{i\epsilon A} d_i \mathrm{tr}|\xi_i(\omega)(U_i(t)\hat{f}(i) - \hat{f}(i))|^2\Big)^{1/2} .$$

We wish to estimate $\Phi(t) = \{\int \sigma_\omega(t)^2\, dP(\omega)\}^{1/2}$. Observe that

$$\Phi^2(t) = \sum_A d_i \mathrm{tr}\; E|\xi_i(U_i(t)\hat{f}(i) - \hat{f}(i))|^2$$

$$= \sum_A d_i E\; \mathrm{tr}(V_i^*(t)\,\xi_i^*\xi_i V_i(t))$$

where, by definition, $V_i(t) = U_i(t)\hat{f}(i) - \hat{f}(i)$. Hence

$$\Phi(t)^2 = \sum_A d_i \mathrm{tr}(V_i^*(t)(E|\xi_i|^2)\, V_i(t))$$

$$= \sum_A d_i \mathrm{tr}(V_i(t)V_i^*(t)\, E|\xi_i|^2)$$

$$\leq \sum_A d_i \| V_i(t) V_i^*(t) E|\xi_i|^2 \|_1$$

$$\leq \sum_A d_i \| V_i(t) V_i^*(t) \|_1 \| E|\xi_i|^2 \|_\infty$$

$$\leq \Big(\sum_A d_i \operatorname{tr} |V_i(t)|^2\Big) \cdot \sup_{i \in A} \| E|\xi_i|^2 \|_\infty .$$

Now if we set $\sigma(t) = \Big(\sum_{i \in A} d_i \operatorname{tr} |U_i(t) \hat{f}(i) - \hat{f}(i)|^2\Big)^{1/2}$ we have that

$$(E\, \sigma_\omega^2(t))^{1/2} \leq \sigma(t) \sup_{i \in A} \| E|\xi_i|^2 \|_\infty^{1/2} .$$

Given this inequality the proof of Lemma 3.6 is completed by using (3.8) and by repeating word for word the argument at the beginning of Chapter III to prove the first part of Theorem 1.1.1 up to Lemma 3.1.1.

We also need a simple technical lemma.

LEMMA 3.7. *Define, for all* $c > 0$,

$$\alpha(c) = \sup_{i \in \Sigma} \| E(|G_i|^2 1_{\{\|G_i\|_\infty > c\}}) \|_\infty^{1/2} .$$

Then $\alpha(c) \to 0$ *when* $c \to \infty$.

Proof. Clearly we have $\alpha(c) \leq \beta(c)$ where $\beta(c)$ is defined by

$$\beta(c) = \sup_{i \in \Sigma} \{ E(\|G_i\|_\infty^2 1_{\{\|G_i\|_\infty > c\}}) \}^{1/2} .$$

Therefore it is enough to show that $\beta(c) \to 0$ when $c \to \infty$. We have by (1.2) and (2.4.5)

$$(E\|G_i\|_\infty^4)^{1/4} \leq \gamma$$

for some absolute constant γ. Therefore, by the Cauchy-Schwarz inequality,

$$\{E(\|G_i\|_\infty^2 1_{\{\|G_i\|_\infty > c\}})\}^{1/2}$$
$$\le (E\|G_i\|_\infty^4)^{1/4}[P(\{\|G_i\|_\infty > c\})]^{1/4}$$
$$\le \gamma[c^{-4}E\|G_i\|_\infty^4]^{1/4}$$
$$\le \gamma^2/c$$

which shows indeed that $\beta(c) \to 0$ when $c \to \infty$.

As in the commutative case, the following lemma is the key step:

LEMMA 3.8. *There is an absolute constant* C_1'' *such that, for any finite subset* A *of* Σ,

$$\left(E\|\sum_{i\in A} d_i \operatorname{tr} G_i U_i \hat{f}(i)\|^2\right)^{1/2} \le C_1''\left(E\|\sum_{i\in A} d_i \operatorname{tr} \varepsilon_i U_i \hat{f}(i)\|^2\right)^{1/2}. \tag{3.13}$$

Proof. We "truncate" G_i as follows: We first choose c large enough so that $\alpha(c) \le 1/2C_3$; (such a c exists by Lemma 3.7). We define G_i' and G_i'' by

$$G_i' = G_i 1_{\{\|G_i\|_\infty \le c\}}$$
$$G_i'' = G_i 1_{\{\|G_i\|_\infty > c\}}.$$

Since $\{\varepsilon_i G_i\}$ and $\{G_i\}$ have the same distribution (recall that we assume $\{\varepsilon_i\}$ and $\{G_i\}$ to be independent of each other), we have

$$E\|\sum_A d_i \operatorname{tr} G_i U_i \hat{f}(i)\|^2 = E\|\sum_A d_i \operatorname{tr} \varepsilon_i G_i U_i \hat{f}(i)\|^2.$$

We write simply $M = \left(E\|\sum_A d_i \operatorname{tr} G_i U_i \hat{f}(i)\|^2\right)^{1/2}$. By the triangle inequality,

$$M \le \left(E\|\sum_A d_i \operatorname{tr} \varepsilon_i G_i' U_i \hat{f}(i)\|^2\right)^{1/2} + \left(E\|\sum_A d_i \operatorname{tr} \varepsilon_i G_i'' U_i \hat{f}(i)\|^2\right)^{1/2}.$$

Hence, using (2.4) (taking into account Remark 2.2) and (3.12),

$$M \le c\left(E\|\sum_A d_i \operatorname{tr} \varepsilon_i U_i \hat{f}(i)\|^2\right)^{1/2} + M\, C_3\, \alpha(c)$$

and by the initial choice of c, we finally obtain

$$M \le 2c\left(E\|\sum_A d_i \operatorname{tr} \varepsilon_i U_i \hat{f}(i)\|^2\right)^{1/2}.$$

This concludes the proof of Lemma 3.8.

Proof of Theorem 3.5. Assume first that $\hat{f}(i) \neq 0$ for at most finitely many i in Σ. In that case, (3.9) follows from (3.13), (2.8) and Corollary 2.12; (3.10) follows from (3.12), (3.13) and Corollary 2.12, and lastly, (3.11) follows from (2.7) in the real case and (2.10) in the complex case. Once this has been checked, all the statements about the a.s. continuity of $X(t)$ or $Y(t)$ follow from Remark 3.3. Moreover, Remark 3.3 also explains why (3.9) and (3.10) are true in general (not only for finite sums) whenever $X(t)$ (or equivalently $Y(t)$) is a.s. continuous. Finally, the statements about the a.s. continuity of $\{Z(t)\}$ can be completed by repeating the same argument as in the commutative case (cf. Chapter III Section 1).

REMARK 3.9. The reader can check easily that if we replace $Z(t)$ by

$$Z'(t) = \sum_{i\in\Sigma} d_i \operatorname{tr} \xi_i \varepsilon_i U_i(t) \hat{f}(i)$$

and if we set $\lambda' = \sup_{i\in\Sigma} \|E(\xi_i \xi_i^*)\|_\infty^{1/2}$ and $\mu' = \sup_{i\in\Sigma} \|(E\sqrt{\xi_i \xi_i^*})^{-1}\|_\infty$ then (iii) of Theorem 3.5 holds with Z' in the place of Z and with λ', μ' instead of λ, μ. It is also possible to study a process of the form

$$Z''(t) = \sum_{i\in\Sigma} d_i \operatorname{tr} \xi_i \varepsilon_i \eta_i U_i(t) \hat{f}(i)$$

where $\{\xi_i\}$ and $\{\eta_i\}$ are two families of random operators independent of $\{\varepsilon_i\}$. If we assume that $\{\xi_i\}$ and $\{\eta_i\}$ are independent of each other, the above method can also be applied. We leave this to the reader.

REMARK 3.10. In [13], it was proved that if a random series of the form (3.3) or (3.4) is a.s. in $L_\infty(G)$, then it is in fact a.s. continuous.

This was originally proved by Billard (cf. [26], p. 49) in the Abelian case. We wish to indicate how this fact can be derived from the preceding results. Assume that (for instance) $\sum_{i\in\Sigma} d_i \operatorname{tr} \varepsilon_i U_i \hat{f}(i)$ represents a.s. a function in $L_\infty(G)$. Then, the following holds:

$$(3.14)\qquad \begin{cases} \forall\varepsilon > 0\ \exists R \geq 0 \text{ such that } \forall F \subset \Sigma,\ F \text{ finite} \\ P\left(\left\{\left\|\sum_{i\in F} d_i \operatorname{tr} \varepsilon_i U_i \hat{f}(i)\right\| > R\right\}\right) < \varepsilon . \end{cases}$$

Indeed, this follows easily either from an argument of Kahane using approximate units (cf. [13], p. 55, for details) or from the proof of Paul Levy's inequality (cf. [26], p. 12). Using Theorem 2.7, we see that (3.14) implies $\exists M \geq 0$ such that $\forall F \subset \Sigma$, F finite,

$$E\left\|\sum_{i\in F} d_i \operatorname{tr} \varepsilon_i U_i \hat{f}(i)\right\| \leq M .$$

Now, if we apply (3.8) and (3.9) we find

$$\sup\{\underline{I}(f_F) \mid F \subset \Sigma, |F| < \infty\} < \infty$$

where we have denoted f_F the function $\sum_{i\in F} d_i \operatorname{tr} U_i \hat{f}(i)$. But clearly, $\underline{I}(f) = \sup\{\underline{I}(f_F) \mid F \subset \Sigma, |F| < \infty\}$, so that, $\underline{I}(f) < \infty$. Therefore, (3.3) is a.s. continuous since, by Theorem 3.4, $\underline{I}(f) < \infty$ is a (necessary and) sufficient condition for the a.s. continuity of (3.3). Of course, there is a similar argument with $\{W_i\}$ in the place of $\{\varepsilon_i\}$. (The result proved in [13] applies as well to more general random series which are not considered in this paper.) In the Gaussian case, results very similar to the above follow from Belyaev [2] (see Theorem 2.3.7).

To summarize the preceding remark, we emphasize: Let $\{F_n\}$ be an increasing sequence of finite subsets of Σ with $\bigcup_n F_n = \Sigma$. Then (3.3) is a.s. continuous if and only if it is a.s. in $L_\infty(G)$ or if and only if (3.14) holds or if and only if

$$\sup_n \|\sum_{i\epsilon F_n} d_i \mathrm{tr}\, \varepsilon_i U_i \hat{f}(i)\| < \infty \text{ a.s.}$$

REMARK 3.11. Using all the preceding results the reader can check easily that the C(G)-valued random variables (3.5) verify the Central Limit Theorem, as soon as

$$\underline{I}(f) < \infty \quad \text{and} \quad \sup_{i\epsilon\Sigma} \|E|\xi_i|^2\|_\infty < \infty .$$

Moreover, if $\sup_{i\epsilon\Sigma} \|(E|\xi_i|^2)^{-1}\|_\infty < \infty$ then $\underline{I}(f) = \infty$ implies that (3.5) is not pre-Gaussian. This shows that the main results of Chapter IV extend in this setting. It is also possible to prove extensions of inequalities (4.1.7), (4.1.10) and of Corollary 4.1.7.

REMARK 3.12. Although the case of randomized Fourier series seems more natural, it is probably useful to underline that all the results of Section 5.3 remain true when $\{U_i\}_{i\epsilon\Sigma}$ is an *arbitrary* collection of finite dimensional representations on G. We have used neither the irreducibility of U_i nor the fact that the representations $\{U_i\}_{i\epsilon\Sigma}$ are mutually non-equivalent.

REMARK 3.13. As the reader may have noticed, we have not yet discussed the case of random Fourier series on a locally compact non-Abelian group. The reason is that in general there is no analogue of the Rademacher and Steinhaus random Fourier series which are the main subject of this paper. Indeed, when the dimension of a representation is infinite, then we can no longer define a "uniformly distributed" random unitary matrix on the corresponding infinite dimensional Hilbert space. However, given a locally compact group G, we can still consider a family $\{U_i\}_{i\epsilon\Sigma}$ of finite dimensional representations on G, and form the corresponding random series as in the beginning of this section. The reader can then check easily (using the same ideas as above) that Theorem 1.1.1 extends to this more general setting. However, this is a rather limited extension, since in general there might not exist any finite dimensional unitary representations on G, except for trivial ones. (This is the case for all non-compact semi-simple Lie groups, such as, for example, $SL(2, R)$.)

CHAPTER VI

APPLICATIONS TO HARMONIC ANALYSIS

1. *The Duality Theorem*

The results of this chapter are due to the second named author (cf. [47], [48]). They are a very natural continuation of the preceding chapters. We present them in the non-Abelian case, which causes no significant extra difficulty, with the exception of Theorem 2.4.

These results can be developed using only Gaussian variables and therefore using, essentially, nothing more than the Dudley-Fernique theorem. Consequently, they are practically independent of Chapter 5. However, Chapter 5 shows that all the statements of Chapter 6 are valid with $\{\varepsilon_i\}_{i\in\Sigma}$ or with $\{W_i\}_{i\in\Sigma}$ in the place of $\{\tilde{G}_i\}_{i\in\Sigma}$. Although this fact is not really used, it seems clear that it will be very useful in the future applications of, for example, Theorem 6.1.1 and Corollary 6.1.5.

In this section, we consider the space of almost surely continuous Gaussian random Fourier series on a compact group. This space can be equipped in a natural way with a Banach space structure. We will obtain a concrete description of the dual space as a space of multipliers and this will yield a new characterization, this time Fourier-theoretical, of almost surely continuous random Fourier series.

Let G be a compact group as above; we continue the previous notation. We define the space $C_{a.s.}(G)$ as the linear space of those f in $L_2(G)$ such that the "randomized" Fourier series of f

$$\tilde{X}(t,\omega) = \sum_{i\in\Sigma} d_i \operatorname{tr} \tilde{G}_i(\omega) U_i(t) \hat{f}(i)$$

represents ω-almost surely a continuous function of t. By a general theorem of Fernique, Landau and Shepp (cf. Theorem 2.3.8 or 2.4.7), we have

$$E\|\sum_{i\in\Sigma} d_i \operatorname{tr} \tilde{G}_i U_i \hat{f}(i)\|_{C(G)} < \infty$$

and we may introduce a norm $[[\cdot]]$ on $C_{a.s.}(G)$ by setting for all $f \in C_{a.s.}(G)$

$$[[f]] = E\|\sum_{i\in\Sigma} d_i \operatorname{tr} \tilde{G}_i U_i \hat{f}(i)\|_{C(G)}\,.$$

It is clear that the space $C_{a.s.}(G)$ equipped with this norm becomes a Banach space. (For clarification, see Theorems 2.4.7 and 2.4.3.) In fact, it is clearly isometric to a closed subspace of the Banach space $L^1(\Omega, \mathcal{F}, \mathbf{P}; C(G))$. Note (cf. Theorem 2.4.7) that if (and only if) f belongs to $C_{a.s.}(G)$ then for any increasing sequence (Σ_n) of finite subsets of Σ such that

$$\cup \Sigma_n \supset \{i \in \Sigma \mid \hat{f}(i) \neq 0\}$$

we have necessarily $[[\sum_{i\notin\Sigma_n} d_i \operatorname{tr} U_i \hat{f}(i)]] \to 0$ when $n \to \infty$. In other words, the "trigonometric polynomials" (i.e. the set of f for which $\hat{f}(i) \neq 0$ for at most finitely many $i \in \Sigma$) form a dense subset of $C_{a.s.}(G)$. We will denote $\mathcal{T}(G)$ the space of all such "trigonometric polynomials."

Of course we could have defined equivalently the space $C_{a.s.}(G)$ in many ways and we could have chosen several equivalent norms. Indeed, the results of the preceding sections show that if in the definition of $C_{a.s.}(G)$ and of $[[\ ]]$ we replace the family $\{\tilde{G}_i\}$ by $\{G_i\}$, by $\{\varepsilon_i\}$ or by $\{W_i\}$, then we obtain the same space and an equivalent norm. The reader will also notice that the choice of the first moment in the definition of $[[\ ]]$ is quite arbitrary, any p-th moment $(1 < p < \infty)$ would lead to an equivalent norm (cf. Corollary 2.4.8).

We need to recall some elementary facts and notation concerning Orlicz functions and Orlicz spaces. By definition, an Orlicz function Φ is an increasing convex function $\Phi : \mathbf{R}_+ \to \mathbf{R}_+$ with $\Phi(0) = 0$. The Orlicz space $L_\Phi(G)$ is defined as the space of all measurable functions $f : G \to \mathbf{C}$

such that $\int \Phi(|f/c|)dm < \infty$ for some $c > 0$. The norm of such a function f in $L_\Phi(G)$ is defined by

$$\|f\|_\Phi = \inf\{c>0 | \int \Phi(|f/c|)dm \leq 1\} .$$

We will consider two concrete cases of Orlicz functions. The function ψ defined for all $t \in \mathbf{R}_+$ by $\psi(t) = \exp t^2 - 1$ and the function ϕ defined for all $t \in R_+$ by $\phi(t) = t(1+\operatorname{Log}(1+t))^{1/2}$. We will keep this notation for ϕ and ψ throughout this section.

The Orlicz space L_ϕ is often referred to in the literature as the class $L\sqrt{\operatorname{Log} L}$. The two functions ϕ and ψ are in duality up to an equivalence of Orlicz functions (see [27], for details on the duality of Orlicz functions). We will use only the fact that the dual of the space $L_\phi(G)$ can be identified with $L_\psi(G)$, the duality being given by the usual pairing: if $f \in L_\psi(G)$, $g \in L_\phi(G)$

$$<f, g> = \int f(x)g(x)dm(x) .$$

The identification of $L_\psi(G)$ with $L_\phi(G)^*$ means that a measurable function f belongs to $L_\psi(G)$ if and only if for all $g \in L_\phi(G)$

$$\int |f(x)g(x)|m(dx) < \infty ;$$

and moreover, that there are absolute constants $\Delta_1 > 0$ and Δ_2 such that

$$\frac{1}{\Delta_1} \|f\|_* \leq \|f\|_\psi \leq \Delta_2 \|f\|_* \tag{1.1}$$

where $\|f\|_*$ is defined by

$$\|f\|_* = \sup\left\{\int fgdm \,|\, g \in L_\phi(G),\ \|g\|_\phi \leq 1\right\} .$$

The main result of this section is the following:

THEOREM 1.1 ([47]). *The space* $C_{a.s.}(G)$ *coincides with the set of elements* f *in* $L_2(G)$ *which admit an expansion of the form*

$$(1.2)\qquad \begin{cases} f = \sum_{n=1}^{\infty} h_n * k_n \\ \text{with } h_n \in L_2(G), k_n \in L_\phi(G) \text{ and } \sum_1^{\infty} \|h_n\|_2 \|k_n\|_\phi < \infty . \end{cases}$$

Moreover, if we denote $|||f|||$ *the infimum of* $\sum_1^{\infty} \|h_n\|_2 \|k_n\|_\phi$ *over all such representations of* f, *then we have*

$$(1.3)\qquad \frac{1}{a_1} |||f||| \leq [[f]] \leq a_2 |||f|||$$

where a_1, a_2 *are absolute constants.*

Let us be more explicit about the notation. In the preceding statement and in the sequel, the symbol $h*k$ means, as usual,

$$h*k(x) = \sum_{i \in \Sigma} d_i \mathrm{tr}\, U_i(x) \hat{k}(i) \hat{h}(i) .$$

We will denote by R_t the right translation operator by t, i.e. $R_t f(x) = f(x+t)$ for any measurable function f.

Until we complete the proof of Theorem 1.1 we will denote by $A_{2,\phi}$ the space of all f in $L_2(G)$ which can be represented as in (1.2); equipped with the norm $|||\ \ |||$. This space is a Banach space.

We also introduce the space $M_{2,\psi}$ (resp. $K_{2,\psi}$) of all bounded (resp. compact) operators from $L_2(G)$ into $L_\psi(G)$ which commute with right translations, i.e. such that $TR_t = R_t T$ for any t in G. By a well-known argument, we can associate to any T in $M_{2,\psi}$ its coefficients $\{\hat{T}(i)\}_{i \in \Sigma}$. These coefficients are characterized by the property that for all $f \in \mathcal{T}(G)$, $\hat{T}(i) \in B(H_i)$ for each $i \in \Sigma$ and

$$Tf = \sum_{i \in \Sigma} d_i \operatorname{tr} U_i \hat{f}(i)\hat{T}(i) ,$$

so that $\widehat{(Tf)}(i) = \hat{f}(i)\hat{T}(i)$. Clearly T is determined by $\{\hat{T}(i)\}_{i\in\Sigma}$ and conversely.

We equip $M_{2,\psi}$ (and its subspace $K_{2,\psi}$) with its natural norm when considered as a space of operators, i.e. we define for all $T \in M_{2,\psi}$

$$\|T\|_{2,\psi} = \sup \{ \|Tf\|_{\psi} | f \in L_2(G), \|f\|_2 \leq 1 \} .$$

As is well known, the space $A_{2,\phi}$ is by construction a predual of $M_{2,\psi}$, the duality being defined for all $T \in M_{2,\psi}$ and all $f \in A_{2,\phi}$ by

$$< T, f > = \sum_{i \in \Sigma} d_i \operatorname{tr} \hat{T}(i)\hat{f}(i) .$$

(For details on this duality theorem, see [29] §5.6, page 180, where the case of multipliers from L_p to L_q is treated in detail; the case of Orlicz spaces is entirely similar.) This means that $T \in M_{2,\psi}$ if and only if $\sum_{i\in\Sigma} d_i |\operatorname{tr} \hat{T}(i)\hat{f}(i)| < \infty$ for all $f \in A_{2,\phi}$, and, moreover, the norm of $M_{2,\psi}$ is equivalent to the dual norm of $A_{2,\phi}$. To be more specific, if we define for $T \in M_{2,\psi}$

$$\|T\|_{\sim} = \sup \{ |< T, f >| \, | f \in A_{2,\phi}, |||f||| \leq 1 \}$$

then we find, by a routine computation,

$$(1.4) \qquad \frac{1}{\Delta_2} \|T\|_{2,\psi} \leq \|T\|_{\sim} \leq \Delta_1 \|T\|_{2,\psi} .$$

We also point out that the set $\mathcal{T}(G)$ of all trigonometric polynomials is a dense linear subspace of $A_{2,\phi}$, so that

$$T \in M_{2,\psi} \text{ iff } \sup \{ |< T, f >| \, | f \in \mathcal{T}(G), |||f||| \leq 1 \} < \infty .$$

We will first show that $M_{2,\psi}$ and $C_{a.s.}(G)$ can be put in duality; this follows from the next result.

THEOREM 1.2. *For all* $f \in C_{a.s.}(G)$ *and all* $T \in M_{2,\psi}$ *we have*

$$\sum_{i \in \Sigma} d_i \operatorname{tr} |\hat{T}(i)\hat{f}(i)| \leq C_0[[f]]\, \|T\|_{2,\psi} \tag{1.5}$$

where C_0 *is an absolute constant.*

Proof. We may as well assume that Σ is countable. Indeed, since the set $A = \{i \in \Sigma \mid \hat{f}(i) \neq 0\}$ is countable we can replace G by the quotient group G/G_0 with $G_0 = \{x \in G \mid U_i(x) = \mathrm{Id}_{H_i}, \forall i \in A\}$. Then (cf. [17], Corollary 28.11), G/G_0 is metrizable and its dual object is countable. (It is the subset of Σ "generated" by A.) This change clearly does not affect the desired estimate (1.5). On the other hand, it is of course enough to prove (1.5) for a dense subset of f in $C_{a.s.}(G)$. Now, since Σ is assumed countable it is easy to see that the set of f in $C_{a.s.}(G)$ such that for all $t, s \in G$

$$\|R_t f - R_s f\|_2 = 0 \implies t = s$$

is dense in $C_{a.s.}(G)$. We will prove (1.5) for such a function $f \in C_{a.s.}(G)$.

Let A_i be a unitary operator on H_i such that $\operatorname{tr} \hat{T}(i)\hat{f}(i) A_i = \operatorname{tr} |\hat{T}(i)\hat{f}(i)|$. Consider the function $g = \sum_{i \in \Sigma} d_i \operatorname{tr} U_i \hat{f}(i) A_i$. Since $\{\tilde{G}_i\}$ and $\{A_i \tilde{G}_i\}$ have the same distribution, $g \in C_{a.s.}(G)$ and $[[g]] = [[f]]$. Let $d(s,t) = \|R_s f - R_t f\|_2$ for $s, t \in G$. By our assumption on f, d is a metric on G. We have

$$\begin{aligned} \|(R_s - R_t) T g\|_\psi &= \|T(R_s - R_t) g\|_\psi \\ &\leq \|T\|_{2,\psi} \|(R_s - R_t) g\|_2 , \end{aligned}$$

and since

$$\|(R_s - R_t) g\|_2 = \|(R_s - R_t) f\|_2 = d(s,t)$$

we have

(1.6) $$\|(R_s - R_t)\,Tg\|_\psi \le \|T\|_{2,\psi}\, d(s,t)\,.$$

Assume $\|T\|_{2,\psi} = 1$. The last inequality above means

$$\int \psi\left(\left|\frac{Tg(x+s) - Tg(x+t)}{d(s,t)}\right|\right) m(dx) \le 1\,.$$

With an additional integration and change of variables, taking into account the fact that $d(x+s, x+t) = d(s,t)$, we get

$$\iint \psi\left(\left|\frac{Tg(s) - Tg(t)}{d(s,t)}\right|\right) m(ds)\, m(dt) \le 1\,.$$

Now since $f \in C_{a.s.}(G)$, we know by Theorem 5.3.4 that $\underline{I}(f) \le C[[f]]$ for some absolute constant C. Since $m(\{x \in G \,|\, d(x,t) < \varepsilon\})$ is independent of t in G, we set

$$m(\varepsilon) = m(\{x \in G \,|\, d(x,0) < \varepsilon\})\,.$$

Clearly (see Remark 2.1.2) we have

$$1 \le m(\varepsilon)\, N(\varepsilon)$$

so that

$$\int_0^\infty (\log 1/m(\varepsilon))^{1/2}\, d\varepsilon \le C[[f]]\,.$$

We may now apply Preston's theorem (Theorem 2.3.9) to the function Tg. This shows that $Tg \in C(G)$ and also (cf. 2.3.36)

$$\|Tg\|_{C(G)} \le \|Tg\|_2 + 20 \int_0^R \psi^{-1}\left(\frac{1}{m\left(\frac{\varepsilon}{2}\right)^2}\right) d\varepsilon\,,$$

with $R = \sup d(s,t) \le 2\|f\|_2$. We have

$$\int_0^R \psi^{-1}\left(\frac{1}{m\left(\frac{\varepsilon}{2}\right)^2}\right)d\varepsilon \le \int_0^R \left(1 + \log \frac{1}{m(\varepsilon/2)^2}\right)^{1/2} d\varepsilon$$

$$\le R + 2\sqrt{2}\int_0^\infty (\log 1/m(\varepsilon))^{1/2} d\varepsilon$$

$$\le 2\|f\|_2 + 2^{3/2} C[[f]] .$$

It is also easy to check that

$$\|Tg\|_2 \le D\|Tg\|_\psi \le D\|T\|_{2,\psi} \|g\|_2 = D\|f\|_2 ,$$

for some absolute constant D. Also, since $\Sigma\, d_i \mathrm{tr}\, \tilde{G}_i \hat{f}(i)$ has the same distribution as $\tilde{g}_1 \cdot \left(\sum d_i \mathrm{tr}\, |\hat{f}(i)|^2\right)^{1/2}$, where $\tilde{g}_1$ is defined on page 6; we have

$$\|f\|_2 = \frac{2}{\sqrt{\pi}} E\left|\sum d_i \mathrm{tr}\, \tilde{G}_i U_i(0) \hat{f}(i)\right|$$

$$\le \frac{2}{\sqrt{\pi}} [[f]] .$$

Therefore, we conclude that

$$\|Tg\|_{C(G)} \le D\|f\|_2 + 20(2\|f\|_2 + 2^{3/2} C[[f]])$$

or

$$\|Tg\|_{C(G)} \le C_0[[f]]$$

where $C_0 = [(2D+80)/\sqrt{\pi}] + 40\sqrt{2}C$. Finally

$$Tg(0) = \sum d_i \mathrm{tr}\, |\hat{T}(i)\, \hat{f}(i)|$$

and since, obviously, $Tg(0) \le \|Tg\|_{C(G)}$ we see that (1.5) follows when $\|T\|_{2,\psi} = 1$. By homogeneity, this is enough to complete the proof of Theorem 1.2.

A fortiori, we have

$$(1.7)\qquad \Big|\sum_{i\in\Sigma} d_i \mathrm{tr}\, \hat{T}(i)\hat{f}(i)\Big| \le C_0[[f]]\|T\|_{2,\psi}$$

for f in $C_{a.s.}(G)$ and T in $M_{2,\psi}$. This allows us to define a duality between $C_{a.s.}(G)$ and $M_{2,\psi}$ by setting for all $T \in M_{2,\psi}$ and all $f \in C_{a.s.}(G)$, $\langle T,f\rangle = \sum d_i \mathrm{tr}\, \hat{T}(i)\hat{f}(i)$. We can rewrite (1.7) as follows: for all $f \in C_{a.s.}(G)$

$$\sup\{|\langle T,f\rangle| \mid T \in M_{2,\psi},\ \|T\|_{2,\psi} \le 1\} \le C_0[[f]]\,.$$

Hence for f in $\mathcal{T}(G)$, we have (cf. (1.4))

$$\sup\{|\langle T,f\rangle| \mid T \in M_{2,\psi},\ \|T\|_{\sim} \le 1\} \le \Delta_2 C_0[[f]]\,.$$

Using the bipolar theorem and the definition of $\|\ \|_{\sim}$ we see that this means simply that for all $f \in \mathcal{T}(G)$

$$(1.8)\qquad |||f||| \le \Delta_2 C_0[[f]]\,.$$

Since $\mathcal{T}(G)$ is dense in $C_{a.s.}(G)$, we have actually that $C_{a.s.}(G) \subset A_{2,\phi}$ and (1.8) holds for any f in $C_{a.s.}(G)$.

It remains to show the reverse inequality. For this we need two easy lemmas; the first one is a simple adaptation of a result of Paley and Zygmund (cf. [26], p. 44).

LEMMA 1.3. *Let* $h \in \mathcal{T}(G)$. *Then*

$$E\Big\|\sum_{i\in\Sigma} d_i \mathrm{tr}\, \tilde{G}_i U_i \hat{h}(i)\Big\|_{\psi} \le \nu\|h\|_2$$

for some absolute constant ν.

Proof. By the integrability properties of scalar Gaussian variables, we have for each fixed x in G,

$$\text{(1.9)} \qquad E\,\psi\left(\left|\frac{\sum d_i \operatorname{tr} \tilde{G}_i U_i(x)\,\hat{h}(i)}{K\|h\|_2}\right|\right) \le 1$$

for some absolute constant K (here we use again the fact that for x fixed, $\sum d_i \operatorname{tr} \tilde{G}_i U_i(x)\hat{h}(i)$ is a complex valued Gaussian random variable with variance equal to $\|h\|_2$). Let us denote $J(\omega)$ the integral

$$J(\omega) = \int \psi\left(\left|\frac{\sum d_i \operatorname{tr} \tilde{G}_i(\omega) U_i(x)\,\hat{h}(i)}{K\|h\|_2}\right|\right) dm(x)\,.$$

By integrating (1.9), we get $EJ \le 1$. We define also

$$N(\omega) = \|\sum d_i \operatorname{tr} \tilde{G}_i(\omega)\, U_i \hat{h}(i)\|_\psi$$

so that by definition of $N(\omega)$, for all $\omega \in \Omega$

$$\text{(1.10)} \qquad \int \psi\left(\left|\frac{\sum d_i \operatorname{tr} \tilde{G}_i(\omega) U_i \hat{h}(i)}{N(\omega)}\right|\right) dm = 1\,.$$

It is easy to check that $a\psi(b) \le \psi(ab)$ for all $a \ge 1$ and all $b \ge 0$. Therefore, if $N(\omega) \ge K\|h\|_2$,

$$\begin{aligned} J(\omega) &\ge \int \frac{N(\omega)}{K\|h\|_2}\, \psi\Big(N(\omega)^{-1} \big|\sum d_i \operatorname{tr} \tilde{G}_i(\omega) U_i \hat{h}(i)\big|\Big) dm \\ &\ge \frac{N(\omega)}{K\|h\|_2}\,, \end{aligned}$$

where, at the last line, we use (1.10). Finally, we have

$$\begin{aligned} EN &\le E(NI_{\{N \le K\|h\|_2\}}) + E(NI_{\{N > K\|h\|_2\}}) \\ &\le K\|h\|_2 + E(K\|h\|_2 J(\omega))\,. \end{aligned}$$

Since $E(J(\omega)) \leq 1$ this is

$$\leq 2K\|h\|_2 .$$

This completes the proof of the lemma.

The second lemma is standard.

LEMMA 1.4. *For all* $k \in L_\phi(G)$ *and all* $k' \in L_\psi(G)$ *we have* $k' * k \in C(G)$ *and* $\|k' * k\|_{C(G)} \leq \Delta_1 \|k\|_\phi \|k'\|_\psi$.

Proof. For all $x, t \in G$ we write $R_x k(t) = k(t+x)$ and also $\check{k}'(t) = k'(-t)$. We have

$$k' * k(x) = \int R_x k(t)\, \check{k}'(t)\, m(dt) = \langle R_x k, \check{k}' \rangle .$$

Since the mapping $x \to R_x k$ is continuous from G into $L_\phi(G)$, and since $L_\psi(G) = L_\phi(G)^*$, $k' * k$ must belong to $C(G)$ and

$$\|k' * k\|_{C(G)} \leq \sup_{x \in G} \|R_x k\|_\phi \cdot \|\check{k}'\|_*$$

$$\leq \Delta_1 \|k\|_\phi \|k'\|_\psi .$$

Proof of Theorem 1.1. Since $\mathcal{T}(G)$ is dense both in $C_{a.s.}(G)$ and in $A_{2,\phi}$ it is enough to show (1.3) for f in $\mathcal{T}(G)$. The left side of (1.3) has already been proved. To prove the remaining part we can assume that f is of the form $f = h * k$ with $k \in L_\phi(G)$, $h \in L_2(G)$ and $h, k \in \mathcal{T}(G)$. We then have

$$\sum d_i \operatorname{tr} \tilde{G}_i U_i \hat{f}(i) = \sum d_i \operatorname{tr} \tilde{G}_i U_i \hat{k}(i) \hat{h}(i)$$

$$= H_\omega * k$$

where H_ω is defined by $H_\omega(x) = \sum d_i \operatorname{tr} U_i(x) \hat{h}(i) \tilde{G}_i(\omega)$. So we have by Lemma 1.4

$$[[f]] = E_\omega \|H_\omega * k\|_{C(G)}$$

$$\leq \Delta_1 \|k\|_\phi E_\omega \|H_\omega\|_\psi .$$

By Lemma 1.3 this is

$$\leq \nu\Delta_1 \|k\|_\phi \|h\|_2 \,.$$

This implies, by convexity, that $[[f]] \leq (\nu\Delta_1)\,|||f|||$ for any f in $\mathcal{T}(G)$ and this completes the proof.

COROLLARY 1.5. *In the duality defined above, we have the identifications:*

$$C_{a.s.}(G)^* = M_{2,\psi}$$

and

$$C_{a.s.}(G) = K_{2,\psi}{}^* \,.$$

Proof. Once we know that $A_{2,\phi}$ can be identified with $C_{a.s.}(G)$ this corollary follows from an argument well known to harmonic analysts.

REMARK 1.6. In the proof of Theorem 6.1.2, for the sake of clarity, we considered only the case when d is a metric. However, this was not really important since the proof of the result of Preston (Theorem 2.3.9) holds even when d is only a pseudo-metric.

REMARK 1.7. It is possible to give an alternate proof of Theorem 6.1.2 using Theorem 2.3.1 instead of Theorem 2.3.9. Indeed (with the same notation as in the above proof) we can apply Theorem 2.3.1 to the process $\{\tilde{X}_t\}_{t\epsilon G}$ defined by $\tilde{X}_t(\omega) = R_t Tg(\omega)$ for all ω in G, equipped with the probability measure m. By (6.1.6) and the discussion preceding Theorem 2.3.1, we find that if $\underline{I}(f) < \infty$ then $\{\tilde{X}_t\}_{t\epsilon G}$ has a version with continuous sample paths. But, of course, this means that Tg itself must be a continuous function on G and the proof can then be completed as above using (2.3.5) instead of (2.3.36).

REMARK 1.8. By Theorem 4.1.5 the space $C_{a.s.}(G)$ is of cotype 2 when G is Abelian. We do not know if this is still true in the non-Abelian case.

REMARK 1.9. There is a way to deduce the main results of Section 5.3 from Theorem 6.1.1. This is indicated in [47] for the Abelian case; the general case is entirely similar.

COROLLARY 1.10. *Let* $F \subset \Sigma$ *be a finite subset of* Σ. *We have*

$$\mathbb{E}\Big\|\sum_{i\in F} d_i \operatorname{tr} \tilde{G}_i U_i\Big\|_{C(G)} \geq \gamma (N \log N)^{1/2} \tag{1.11}$$

where $N = \sum_{i\in F} d_i^2 \geq 3$ *and* $\gamma > 0$ *is an absolute constant.*

Proof. We first claim that for all $p > 2$ and all $f \in L_2(G)$

$$\Big\|\sum_{i\in F} d_i \operatorname{tr} U_i \hat{f}(i)\Big\|_p \leq N^{1/2-1/p} \|f\|_2 .$$

Indeed, this inequality is obtained easily by interpolating between the cases $p = 2$ and $p = \infty$. It follows that for all $p > 2$

$$\Big\|\sum d_i \operatorname{tr} U_i \hat{f}(i)\Big\|_p \leq \alpha (N/\operatorname{Log} N)^{1/2} \sqrt{p}\, \|f\|_2 ,$$

where $\alpha = (2e)^{-1/2}$ (indeed $\sup_{p>2} N^{-1/p} p^{-1/2} = \alpha (\log N)^{-1/2}$. By Lemma 5.2.11, this means that if we define $T \in M_{2,\psi}$ by $\hat{T}(i) = \mathrm{Id}_{H_i}$ if $i \in F$ and $\hat{T}(i) = 0$ otherwise, then we have

$$\|T\|_{2,\psi} \leq \beta (N/\log N)^{1/2}$$

for some constant β.

Applying Theorem 1.2, we find that for all $f \in C_{a.s.}(G)$

$$C_0 [[f]] \, \|T\|_{2,\psi} \geq \sum_{i\in F} d_i \operatorname{tr} |\hat{f}(i)| .$$

Hence

$$[[f]] \geq (1/\beta C_0)(N \log N)^{1/2} \Big(\frac{1}{N} \sum_{i\in F} d_i \operatorname{tr} |\hat{f}(i)|\Big) ,$$

so that finally, taking $f = \sum_{i\in F} d_i \operatorname{tr} U_i$, we obtain (1.11) with $\gamma = (\beta C_0)^{-1}$.

REMARK 1.11. In the particular case when G is the circle group, using (5.3.9) and Remark 5.2.16, we have the following inequality:

For any sequence of integers $k_1 < k_2 < \cdots < k_N$ (with $N \geq 3$)

$$(1.12) \qquad E\left\| \sum_{n=1}^{N} \varepsilon_n e^{ik_n t} \right\|_{C([0,2\pi])} \geq \gamma'(N \log N)^{1/2}$$

for some absolute constant $\gamma' > 0$ (where $\{\varepsilon_n\}$ is a Rademacher or Bernoulli sequence).

It follows from a result of Salem and Zygmund, that when the integers $k_1, k_2, \cdots, k_N$ are consecutive the estimate (1.12) cannot be improved, (cf. [26], p. 55). The inequality (1.12) seems closely related to Theorem 4.5 in [55]. Note that (1.12) also follows from Remark 7.1.3.

2. *Applications to Sidon sets*

In this section, we indicate briefly an application of the duality theorem to Sidon sets. In fact, it is this application which suggested the investigation of the dual of $C_{a.s.}(G)$. We recall first a standard notation and a definition. Let Λ be a subset of Σ. Given a linear space $\mathcal{F}(G)$ of integrable functions on G (for example: $C(G)$, $\mathcal{T}(G)$, $L_p(G)$, $C_{a.s.}(G)$, etc.), we denote $\mathcal{F}(G)_\Lambda$ the subspace of $\mathcal{F}(G)$ spanned by all the functions f in $\mathcal{F}(G)$ such that for all $i \notin \Lambda$, $\hat{f}(i) = 0$.

DEFINITION 2.1. A subset Λ of Σ is called a Sidon set, if there exists a constant χ such that for all $f \in C(G)_\Lambda$

$$\sum_{i \in \Lambda} d_i \operatorname{tr} |\hat{f}(i)| \leq \chi \|f\|_{C(G)} .$$

For details and references see [17], [54] and especially [31]. The next theorem is due to Rudin [53] in the Abelian case and to Figa-Talamanca and Rider ([11], [12], [52]) in the non-Abelian case.

THEOREM 2.2. *Let* $\Lambda \subset \Sigma$ *be a Sidon set. Then*

$$(2.1)\qquad \begin{cases} L_2(G)_\Lambda \subset L_\psi(G)_\Lambda \text{ and there is constant } D \text{ such that} \\ \text{for all } f \in L_2(G)_\Lambda,\ \|f\|_\psi \le D\|f\|_2 . \end{cases}$$

In [53], Rudin raised the question whether the converse is true, i.e. whether (2.1) implies that Λ is a Sidon set. For some particular cases, a positive answer follows from [32]. The general case (especially the case of the circle group) was settled affirmatively in [48].

THEOREM 2.3. *Any set* $\Lambda \subset \Sigma$ *which satisfies (2.1) is necessarily a Sidon set.*

The following result, essentially due to Rider [51], is crucial in the proof.

THEOREM 2.4. *Assume that a subset* Λ *of* Σ *has the following property:*

$$(2.2)\qquad \begin{cases} \text{There is a constant } B \text{ such that} \\ \text{for all } f \in C_{a.s.}(G)_\Lambda,\ \sum_{i\in\Lambda} d_i \operatorname{tr}|\hat{f}(i)| \subseteq B[[f]] . \end{cases}$$

Then Λ *is a Sidon set.*

This theorem is a variant of Rider's result. Actually, Rider uses $\{W_i\}_{i\in\Sigma}$ in place of $\{\tilde{G}_i\}_{i\in\Sigma}$. As we have seen in Chapter V, this merely transforms $[[\ \ ]]$ into an equivalent norm and therefore does not affect the validity of Theorem 2.4. In fact, there is a very simple way (indicated in [48]) to deduce Theorem 2.4 directly from Rider's statement. In his paper [51], Rider proves his version of Theorem 2.4 in the Abelian case only. However, he kindly communicated to us an unpublished proof of the corresponding statement in the non-Abelian case. The proof of Theorem 2.3 is now very short using the duality theorem.

Proof of Theorem 2.3. Consider the orthogonal projection P from $L_2(G)$ onto $L_2(G)_\Lambda$. Assume that Λ satisfies (2.1). Then, we have for all $f \in L_2(G)$

$$\|Pf\|_\psi \leq D\|Pf\|_2 \leq D\|f\|_2 \,.$$

In other words, $P \in M_{2,\psi}$ and $\|P\|_{2,\psi} \leq D$. Clearly, $\hat{P}(i)$ is the identity on H_i for i in Λ and $\hat{P}(i) = 0$ otherwise.

Therefore, applying (1.5), we find that for all $f \in C_{a.s.}(G)_\Lambda$

$$\sum_{i \in \Lambda} d_i \mathrm{tr}\, |\hat{f}(i)| \leq C_0[[f]]\, \|P\|_{2,\psi}$$
$$\leq C_0 D[[f]] \,.$$

We see by Theorem 2.4, that Λ must be a Sidon set.

REMARK 2.5. Applying (5.2.8) in a suitable Banach space, it is easy to see that the converse of Theorem 2.4 is true. It is also easy to see that (2.2) equivalently means that $C_{a.s.}(G)_\Lambda \subseteq C(G)_\Lambda$.

REMARK 2.6. Clearly (2.1) is equivalent to:

$$\begin{cases} L_2(G)_\Lambda \subset \bigcap_{\infty > p > 2} L_p(G)_\Lambda \text{ and there exists a constant } K \\ \text{such that for all } f \in L_2(G)_\Lambda,\ \|f\|_p \leq K\sqrt{p}\,\|f\|_2 \end{cases}$$

(see Lemma 5.2.11).

REMARK 2.7. It was proved by Figa-Talamanca and Rider that if $\{A_i\}$ is an arbitrary collection of operators with $A_i \in B(H_i)$ for each $i \in \Sigma$, then for any $p > 2$

$$\text{(2.3)} \qquad \left(E\left|\sum d_i \mathrm{tr}\, W_i A_i\right|^p\right)^{1/p} \leq C\sqrt{p}\left(\sum d_i \mathrm{tr}\, |A_i|^2\right)^{1/2}$$

for some numerical constant C. (A similar result holds with $\{\varepsilon_i\}$ in the place of $\{W_i\}$.) This result is the main ingredient for the proof of Theorem 2.2 in the non-Abelian case. Its proof (Theorem 36.2 in [17]) is rather long, so we might as well mention that a very simple proof can be obtained using Gaussian variables. Indeed, by the simple Corollary 5.2.4, we have

$$\left(E\left|\sum d_i \mathrm{tr}\, W_i A_i\right|^p\right)^{1/p} \leq (1/\delta)\left(E\left|\sum d_i \mathrm{tr}\, \tilde{G}_i A_i\right|^p\right)^{1/p} .$$

Now let $\tilde{g}$ be a fixed complex Gaussian variable with mean zero and $E|\tilde{g}|^2 = 1$. Since the distribution of $\sum d_i \operatorname{tr} \tilde{G}_i A_i$ is the same as that of $\tilde{g}\left(\sum d_i \operatorname{tr} |A_i|^2\right)^{1/2}$, we have

$$\left(E\left|\sum d_i \operatorname{tr} \tilde{G}_i A_i\right|^p\right)^{1/p} = (E|\tilde{g}|^p)^{1/p}\left(\sum d_i \operatorname{tr} |A_i|^2\right)^{1/2}$$

and (as is classical) we know that

$$(E|\tilde{g}|^p)^{1/p} \in O(\sqrt{p}) .$$

Therefore (2.3) follows. Of course, (2.3) is also a particular case of Corollary 5.2.12, but in this special case the direct argument above is much simpler.

REMARK 2.8. In [49] it is proved that $C_{a.s.}(G) \cap C(G)$ is a Banach algebra for pointwise multiplication and enjoys several other remarkable properties. For more recent results on Sidon sets, see [58].

CHAPTER VII
ADDITIONAL RESULTS AND COMMENTS

1. *Derivation of classical results*

Let us consider random Fourier series on the circle group,

$$(1.1) \qquad Z(t) = \sum_{n=0}^{\infty} a_n \varepsilon_n \xi_n e^{int}, \qquad t \in [0, 2\pi]$$

where $\{\xi_n\}$ satisfies (1.1.10). There are three results that were mentioned in Chapter I, two of them classical and the other obtained by classical methods, that give criteria for the uniform convergence a.s. of the series (1.1). To repeat (see (1.1.5)) we have, for $\{a_n\}$ real,

$$(1.2) \qquad \sum_{n=2}^{\infty} \frac{\left(\sum_{k=n}^{\infty} a_k^2\right)^{1/2}}{n(\log n)^{1/2}} < \infty$$

is a sufficient condition for the uniform convergence a.s. of (1.1) and is also necessary when $|a_n|$ is non-increasing. Also

$$(1.3) \qquad \sum_{n=0}^{\infty} \left(\sum_{j=2^n}^{2^{n+1}-1} a_j^2 \right)^{1/2} = \infty$$

is a necessary condition for the uniform convergence a.s. of (1.1) (see (1.1.3)). (Of course these results were proved for the series (1.1.1) or (1.1.4) but we know from Theorems 1.1.1 or 1.1.5 that they apply to the series (1.1) above.) We will now derive these three results and some interesting generalizations.

The sufficiency of (1.2) follows from the following lemma which extends the result to random Fourier series on R^N. The proof is elementary. It is based on Theorem 1 [33]. Consider

$$\sum_{k=0}^{\infty} a_k \varepsilon_k \xi_k e^{i\langle \lambda_k, t\rangle}, \qquad t \in [-1/2, 1/2]^N \tag{1.4}$$

where $\{\lambda_k\}$ is a sequence of elements in R^N and $\{a_k\}$, $\{\varepsilon_k\}$ and $\{\xi_k\}$ are as in (1.1). Let

$$S(\{a_k\},\{\lambda_k\}) = \sum_{n=2}^{\infty} \frac{\left(\sum_{|\lambda_k|\geq n} |a_k|^2\right)^{1/2}}{n(\log n)^{1/2}}$$

where $|\cdot|$ denotes the Euclidean norm. Then for $\overline{\sigma}(u)$ defined as in Chapter I, we have

LEMMA 1.1:

$$\int_0^{2^N} \frac{\overline{\sigma}(u)}{u\left(\log \frac{4^N}{u}\right)^{1/2}}\, du \leq C(N)\left[\left(\sum_k |a_k|^2\right)^{1/2} + S(\{a_k\},\{\lambda_k\})\right] \tag{1.5}$$

where $C(N)$ *is a constant depending on* N.

Proof. Let $\hat{\sigma}(d) = \sup_{|u|\leq d} \sigma(u)$ and recall (see (1.1.12)) that the domain of $\sigma(u)$ is $[-1,1]^N$. It follows that for $0 \leq d \leq 1$, $\hat{\sigma}(d) \geq \overline{\sigma}(B_N d^N)$ for some constant $B_N > 0$ depending on N. Note that we may assume that $B_N/2^N \leq 1$. We have

$$\int_0^{B_N/2^N} \frac{\overline{\sigma}(u)}{u\left(\log \frac{4^N}{u}\right)^{1/2}}\, du \leq \int_0^{B_N/2^N} \frac{\hat{\sigma}\left(\left(\frac{u}{B_N}\right)^{1/N}\right)}{u\left(\log \frac{4^N}{u}\right)^{1/2}}\, du \tag{1.6}$$

$$\leq \sqrt{N} \int_0^{1/2} \frac{\hat{\sigma}(s)}{s(\log 1/s)^{1/2}}\, ds .$$

Also, clearly,

$$\int_{B_N/2^N}^{2^N} \frac{\overline{\sigma}(u)}{u\left(\log \frac{4^N}{u}\right)^{1/2}}\, du \le C'(N)\left(\sum_{k=0}^{\infty} |a_k|^2\right)^{1/2} .$$

where $C'(N)$ is a constant depending on N. Thus to obtain (1.5) we need to consider the final integral in (1.6). Note that

$$(1.7) \qquad (\log 2)^{-1/2} \int_0^{1/2} \frac{\hat{\sigma}(s)}{s(\log 1/s)^{1/2}}\, ds \le \sum_{n=1}^{\infty} \frac{\hat{\sigma}(2^{-n})}{n^{1/2}} .$$

Also

$$\hat{\sigma}(2^{-n}) = 2 \sup_{|u|\le 2^{-n}} \left(\sum_{k=0}^{\infty} |a_k|^2 \sin^2 \frac{<\lambda_k, u>}{2}\right)^{1/2}$$

$$\le 2\left[2^{-n}\left(\sum_{|\lambda_k|<2^n} |a_k|^2|\lambda_k|^2\right)^{1/2} + \left(\sum_{|\lambda_k|\ge 2^n} |a_k|^2\right)^{1/2}\right].$$

Let

$$t_j = \left(\sum_{2^j\le|\lambda_k|<2^{j+1}} |a_k|^2\right)^{1/2}, \qquad j \ge 1 .$$

Then for $n \ge 1$, we have

$$(1.8) \qquad 4^{-1}\hat{\sigma}(2^{-n}) \le 2^{-n}\left(\sum_{k=0}^{\infty} |a_k|^2\right)^{1/2} + 2^{-n}\sum_{j=1}^{n-1} t_j 2^j$$

$$+ \left(\sum_{j=n}^{\infty} t_j^2\right)^{1/2}$$

and therefore

$$4^{-1}\sum_{n=1}^{\infty}\frac{\hat{\sigma}(2^{-n})}{n^{1/2}} \le \Big(\sum_{k=0}^{\infty}|a_k|^2\Big)^{1/2} + \sum_{n=1}^{\infty}2^{-n}\sum_{j=1}^{n-1}t_j 2^j$$

$$+\sum_{n=1}^{\infty}\Big(\frac{1}{n}\sum_{j=n}^{\infty}t_j^2\Big)^{1/2}.$$

Also

$$\sum_{n=1}^{\infty}2^{-n}\sum_{j=1}^{n-1}t_j 2^j \le \sum_{j=1}^{\infty}t_j \le 2\sum_{n=1}^{\infty}\Big(\frac{1}{n}\sum_{j=n}^{\infty}t_j^2\Big)^{1/2},$$

where at the last step we use a well-known inequality (see [16] Theorem 345 or [25], page 146). Since

$$\sum_{n=1}^{\infty}\Big(\frac{1}{n}\sum_{j=n}^{\infty}t_j^2\Big)^{1/2} \le 8S(\{a_k\},\{\lambda_k\})$$

we get (1.5).

If we use Lemma 1.1 in Theorem 1.1.1 or 1.1.4 we see that $S(\{a_k\},\{\lambda_k\}) < \infty$ is sufficient for the uniform convergence a.s. of the series (1.4) and hence also (1.1).

When $|a_n|$ is non-increasing (1.2) is also a necessary condition for the uniform convergence a.s. of (1.1). This result was proved in [35]. It was obtained for Gaussian random Fourier series in [34]. The proof is much simpler in the Gaussian case because we can use Slepian's lemma. Since we now know that the Gaussian and Rademacher series are equivalent the simpler proof suffices. We will also see that this result follows immediately from the next theorem which is also valid for random Fourier series on compact Abelian groups.

As in (1.1.11) consider the series

(1.9) $$Z(x) = \sum_{\gamma\epsilon\Gamma} a_\gamma \epsilon_\gamma \xi_\gamma \gamma(x), \qquad x \epsilon G$$

where G is a compact group with dual group Γ and $\{\xi_\gamma\}$ satisfies (1.1.10). Following (1.1.12) we consider the translation invariant metric

$$\sigma(u) = \Big(\sum_{\gamma\in\Gamma} |a_\gamma|^2 |\gamma(u)-1|^2\Big)^{1/2} \tag{1.10}$$

and $\overline{\sigma}(u)$ the non-decreasing rearrangement of σ. We have the following lower bound for $\overline{\sigma}$.

THEOREM 1.2. *For $\overline{\sigma}$ as defined above, where G is a compact Abelian group, and for integers $n \geq 1$*

$$2\overline{\sigma}(1/n) \geq \Big(\sum_{k>n}^{\infty} (a_k^*)^2\Big)^{1/2}, \tag{1.11}$$

where $\{a_k^*\}$ is the non-increasing rearrangement of the numbers $\{|a_\gamma|\}_{\gamma\in\Gamma}$.

Proof. Consider the operator $T: L_2(G) \to C(G)$ defined for $\xi \in L_2(G)$ by

$$T\xi(x) = \int_G \xi(t)\overline{f_x}(t)\, m(dt)$$

where $f(t) = \sum a_\gamma \gamma(t)$, $f_x(t) = f(t+x)$ and m is the Haar measure on G. We define, as is standard, the approximation number

$$a_n(T) = \inf\{\|T-S\|\}$$

where the infimum is over all operators $S: L_2(G) \to C(G)$ of rank less than n. By Remark 2.1.2 we have,

$$N_\sigma(G, 2(\overline{\sigma}(1/n)+\delta)) \leq m_\sigma^{-1}(\overline{\sigma}(1/n)) \leq n .$$

This implies that there exist $t_1, \cdots, t_n \in G$ such that

$$\sup_t \inf_{1\leq j\leq n} \sigma(t-t_j) \leq 2\overline{\sigma}(1/n) . \tag{1.12}$$

Let $P : L_2(G) \to L_2(G)$ be the orthogonal projection onto the span of $f_{t_1}, \cdots, f_{t_n}$. The first step of the proof is to show that

$$\|T - TP\| \leq 2\bar{\sigma}(1/n) . \tag{1.13}$$

Note that $T\xi(x) = \langle f_x, \xi \rangle$ where $\langle \cdot, \cdot \rangle$ is the inner product in $L^2(G)$. Therefore

$$(T\xi - TP\xi)(x) = \langle (I-P)f_x, \xi \rangle$$

and hence

$$\|(T-TP)\xi\|_{C(G)} = \sup_x |\langle (I-P)f_x, \xi \rangle|$$

$$\leq \sup_x \|(I-P)f_x\|_2 \|\xi\|_2 .$$

Therefore, the operator norm

$$\|T - TP\| \leq \sup_x \|(I-P)f_x\|_2 . \tag{1.14}$$

Note that $\|f_x - f_{t_j}\|_2 = \sigma(x - t_j)$. Therefore, by (1.12),

$$\sup_{x \in G} \inf_{1 \leq j \leq n} \|f_x - f_{t_j}\|_2 \leq 2\bar{\sigma}(1/n)$$

and consequently for each $x \in G$ there exists a $t_j \in \{t_1, \cdots, t_n\}$ such that

$$\|(I-P)f_x\|_2 = \|(I-P)(f_x - f_{t_j})\|_2 \leq 2\bar{\sigma}(1/n) . \tag{1.15}$$

It follows from (1.14) and (1.15) that (1.13) is valid.

Let $j : C(G) \to L^2(G)$ be the canonical injection. Note that if $B : L^2(G) \to C(G)$ is a bounded linear operator then $jB : L^2(G) \to L^2(G)$ is a Hilbert-Schmidt operator and

$$\|jB\|_{HS} \leq \|B\| \tag{1.16}$$

where $\| \quad \|_{HS}$ denotes the Hilbert-Schmidt norm. This is a well known fact but we will include a proof for the sake of completeness. Since

$\{\gamma | \gamma \epsilon \Gamma\}$ is an orthonormal basis of $L_2(G)$ we have

$$\|jB\|_{HS} = \Big(\sum_{\gamma} \|jB(\gamma)\|^2_{L^2(G)}\Big)^{\frac{1}{2}} . \tag{1.17}$$

Let $\phi_\gamma(t) = B(\gamma)(t)$ so that (1.17) can be written as

$$\Big(\sum_{\gamma} \int_G |\phi_\gamma(t)|^2 m(dt)\Big)^{\frac{1}{2}} . \tag{1.18}$$

Note that

$$\begin{aligned}\sup_t \Big|\sum_{\gamma} a_\gamma \phi_\gamma(t)\Big| &= \Big\|\sum_{\gamma} a_\gamma B(\gamma)\Big\|_{C(G)} \\ &\le \Big(\sum_{\gamma} |a_\gamma|^2\Big)^{\frac{1}{2}} \|B\| ,\end{aligned}$$

which implies that

$$\Big(\sum_{\gamma} |\phi_\gamma(t)|^2\Big)^{\frac{1}{2}} \le \|B\| . \tag{1.19}$$

Substituting (1.19) in (1.18) we get (1.16).

Let $A = jT$. Then by (1.16) we have

$$\|A-AP\|_{HS} = \|j(T-TP)\|_{HS} \le \|T-TP\| . \tag{1.20}$$

We will show that

$$\|A-AP\|_{HS} \ge \Big(\sum_{k>n}^{\infty} (a_k^*)^2\Big)^{\frac{1}{2}} . \tag{1.21}$$

This last inequality combined with (1.20) and (1.13) will complete the proof of the theorem.

We will now establish (1.21).

$$\begin{aligned}\|A-AP\|_{HS} &= (\mathrm{tr}(A(I-P))^*A(I-P))^{\frac{1}{2}} \\ &= (\mathrm{tr}(I-P)\,A^*A(I-P))^{\frac{1}{2}} \\ &= (\mathrm{tr}|A|^2(I-P))^{\frac{1}{2}} \\ &= (\mathrm{tr}|A|^2 - \mathrm{tr}|AP|^2)^{\frac{1}{2}} .\end{aligned} \tag{1.22}$$

It is well known that

$$\text{(1.23)} \qquad \operatorname{tr}|A|^2 = \|A\|_{HS}^2 = \sum_{k=1}^{\infty} a_k(A)^2 .$$

Also, since rank $AP \leq n$, $a_{n+1}(AP) = 0$. Hence

$$\text{(1.24)} \qquad \operatorname{tr}|AP|^2 = \sum_{k=1}^{\infty} a_k(AP)^2 \leq \sum_{k=1}^{n} a_k(A)^2 ,$$

where the last inequality follows because, obviously, $a_k(AP) \leq a_k(A)$. Combining (1.22), (1.23) and (1.24) we get

$$\text{(1.25)} \qquad \|A-AP\|_{HS} \geq \Big(\sum_{k=n+1}^{\infty} a_k^2(A) \Big)^{1/2} .$$

Finally, we observe that

$$A\xi(x) = \sum_{\gamma \epsilon \Gamma} \hat{\xi}(\gamma)\, \overline{\hat{f}(\gamma)}\, \overline{\gamma(x)}$$

where $\{\hat{\xi}(\gamma)\}$ are the Fourier coefficients of ξ. Therefore, by a well-known result (see e.g. Dunford and Schwartz, Linear Operators, Vol. 2, Lemma XI, 9.2) we have

$$\text{(1.26)} \qquad a_k(A) = a_k^* .$$

Substituting (1.26) into (1.25) we get (1.21).

REMARK 1.3. By Theorem 1.1.4, with $Z(x)$ as given in (1.9) we have

$$\text{(1.27)} \qquad E\|Z(x)\|_{C(G)} \geq C \sum_{n=2}^{\infty} \frac{\Big(\sum_{k=n}^{\infty} (a_k^*)^2 \Big)^{1/2}}{n(\log n)^{1/2}} \inf_{\gamma} E|\xi_\gamma|$$

for some absolute constant C. Thus we obtain that (1.2) is necessary for the uniform convergence a.s. of (1.1) when $|a_n|$ is non-increasing. We can also use (1.27) to obtain (6.1.12). In fact, (1.27) along with Theorem 1.1.4 and Lemma 1.1 implies the following curious result.

THEOREM 1.4. *If* $\sum_{\gamma\epsilon\Gamma} a_\gamma \varepsilon_\gamma \xi_\gamma \gamma(x), x \epsilon G$, *converges uniformly a.s. then so does* $\sum_{k=0}^{\infty} a_k^* \varepsilon_k e^{ikt}, t \epsilon [0, 2\pi]$.

REMARK 1.5. If G is a non-Abelian compact group, then consider $A : L_2(G) \to L_2(G)$ defined by $A(\xi) = \xi * f$ where f is a fixed function in $C_{a.s.}(G)$. Consider (with the notation of Chapter V) the Fourier coefficients $\{\hat{f}(i) | i \epsilon \Sigma\}$ of f. For each i in Σ, we denote $\{\lambda_j^i | 1 \le j \le d_i\}$ the eigenvalues of $|\hat{f}(i)| = (\hat{f}(i)^* \hat{f}(i))^{1/2}$ repeated according to their multiplicity. Now we define $\{a_n^*\}$ as the decreasing rearrangement of the collection $\{\lambda_j^i | 1 \le j \le d_i, i \epsilon \Sigma\}$ where each λ_j^i is repeated d_i times. With these conventions, it is easy to check that $a_n(A) = a_n^*$ for each $n \ge 0$. Therefore, by the same argument as above, if f belongs to $C_{a.s.}(G)$, then necessarily

$$\sum_{n>1} \frac{\left(\sum_{k>n} a_n^{*2}\right)^{1/2}}{n(\log n)^{1/2}} < \infty .$$

We conclude this section with a derivation of the classical Paley-Zygmund result that (1.3) is a necessary condition for the uniform convergence a.s. of (1.1). Actually our proof yields a somewhat more general result formulated for an arbitrary compact Abelian group in terms of what we call a "Sidon partition." (What follows also immediately extends to the non-Abelian case but, for simplicity, we will state our results in the Abelian case only.)

Let G be a compact Abelian group with dual group Γ. Let $\{\Gamma_j\}_{j\epsilon J}$ be a disjoint partition of Γ. We will call $\{\Gamma_j\}_{j\epsilon J}$ a Sidon partition if there is a constant K such that any set $\{\gamma_j | j \epsilon J\}$ with $\gamma_j \epsilon \Gamma_j$ for each $j \epsilon J$ is a Sidon set with Sidon constant less than K. With this notation we have the following:

THEOREM 1.6. *Assume that* $\sum_{\gamma\epsilon\Gamma} a_\gamma \varepsilon_\gamma \gamma(t), t \epsilon G$ *converges uniformly a.s. Then, for any Sidon partition* $\{\Gamma_j\}_{j\epsilon J}$ *of* Γ, *we must have*

$$\sum_{j\epsilon J}\Big(\sum_{\gamma\epsilon\Gamma_j} |a_\gamma|^2\Big)^{1/2} < \infty .$$

This theorem is a consequence of the following lemma which follows from a classical result of Zygmund.

LEMMA 1.7. *Let* $\{\Gamma_j\}_{j\epsilon J}$ *be a Sidon partition of* Γ *with associated constant* K. *Then*

(1.28)
$$\|\sum_{\gamma\epsilon\Gamma} a_\gamma \gamma\|_\psi \le C\Big(\sum_{j\epsilon J}\Big(\sum_{\gamma\epsilon\Gamma_j} |a_\gamma|\Big)^2\Big)^{1/2}$$

where $C = C(K)$ *depends only on* K. *(Recall that* $\|\ \|_\psi$ *is the Orlicz space norm associated with the function* $\psi(x) = \exp x^2 - 1$ *as defined in Chapter VI, Section 1.)*

Proof. For any set $\{\gamma_j | j \epsilon J\}$ with $\gamma_j \epsilon \Gamma_j$ for each $j \epsilon J$ we have ,

(1.29)
$$\|\sum_j \lambda_j \gamma_j\|_\psi \le C\Big(\sum_j |\lambda_j|^2\Big)^{1/2}$$

where $C = C(K)$ depends only on K. This is a generalization of a classical result of Zygmund and can be derived by standard techniques from 5.7.7 [54]. We next show that (1.29) implies that if $a_\gamma \ge 0$ and $\sum_{\gamma\epsilon\Gamma_j} a_\gamma = 1$ for all $j \epsilon J$ then

(1.30)
$$\|\sum_{j\epsilon J} \lambda_j \sum_{\gamma\epsilon\Gamma_j} a_\gamma \gamma\|_\psi \le C\Big(\sum |\lambda_j|^2\Big)^{1/2} .$$

It is clear that by homogeneity (1.30) implies (1.28). Therefore it is enough to prove (1.30). The latter inequality follows by a simple convexity argument. Indeed, consider the probability measure ν_j on Γ_j defined by $\nu_j(\{\gamma\}) = a_\gamma$ for all $\gamma \epsilon \Gamma_j$, and define the probability measure ν on $\prod_{j=1}^{\infty} \Gamma_j$ by $\nu = \bigotimes_{j=1}^{\infty} \nu_j$. We can observe that $\sum_{\gamma \epsilon \Gamma_j} a_\gamma \gamma = \int \gamma_j d\nu_j(\gamma_j)$ and therefore

$$\sum_{j \epsilon J} \lambda_j \Big(\sum_{\gamma \epsilon \Gamma_j} a_\gamma \gamma\Big) = \int \Big(\sum \lambda_j \gamma_j\Big) d\nu . \tag{1.31}$$

Since $\|\cdot\|_\psi$ is a norm, it follows from (1.31) that

$$\Big\|\sum \lambda_j \Big(\sum_{\gamma \epsilon \Gamma_j} a_\gamma \gamma\Big)\Big\|_\psi \le \int \Big\|\sum \lambda_j \gamma_j\Big\|_\psi \, d\nu$$

$$\le C\Big(\sum |\lambda_j|^2\Big)^{1/2},$$

where we use (1.29) for the second inequality. This concludes the proof of (1.30) and of Lemma 1.7.

Proof of Theorem 1.6. Assume that the Gaussian random Fourier series $\sum_{\gamma \epsilon \Gamma} a_\gamma g_\gamma \gamma(t)$, $t \epsilon G$ converges uniformly a.s. Then by Theorem 1.1.1 and Lemma 2.3.6 we have that the entropy integral $J(G, d) < \infty$ where $d(s, t) = \Big(\sum_{\gamma \epsilon \Gamma} |a_\gamma|^2 |\gamma(s) - \gamma(t)|^2\Big)^{1/2}$. Now let us consider the series

$$\tilde{X}_t(\omega) = \sum_{j \epsilon J} \sum_{\gamma \epsilon \Gamma_j} |a_\gamma b_\gamma| \gamma(t + \omega)$$

where $\{\Gamma_j\}$ is the same as above, $\omega, t \epsilon G$ and $\{b_\gamma\}$ are chosen such that

$$\sum_{\gamma \epsilon \Gamma_j} |b_\gamma|^2 \le 1 \quad \text{and} \quad \sum_{\gamma \epsilon \Gamma_j} |a_\gamma b_\gamma| = \Big(\sum_{\gamma \epsilon \Gamma_j} |a_\gamma|^2\Big)^{1/2} .$$

We can now easily check that

$$(1.32)\quad E_\omega \exp\left\{\frac{|\tilde{X}_t - \tilde{X}_s|}{Cd(s,t)}\right\}^2 \le E_\omega \exp\left\{\frac{|\tilde{X}_t - \tilde{X}_s|^2}{C^2 \sum_j \left(\sum_{\gamma\epsilon\Gamma_j} |a_\gamma b_\gamma|\,|\gamma(s)-\gamma(t)|\right)^2}\right\}$$

where E_ω denotes integration in the ω variable with respect to Haar measure on G.

Note that for fixed $s, t \epsilon G$

$$\tilde{X}_t - \tilde{X}_s = \sum_{j\epsilon J} \sum_{\gamma\epsilon\Gamma_j} |a_\gamma b_\gamma|(\gamma(t)-\gamma(s))\gamma(\omega), \qquad \omega \epsilon G$$

is a Fourier series on G. It follows by Lemma 1.7 that the right side of (1.32) is less than or equal to 2. Hence

$$E_\omega \exp\left\{\frac{|\tilde{X}_t - \tilde{X}_s|}{Cd(s,t)}\right\}^2 \le 2 .$$

This last inequality implies that

$$P\left\{\frac{|\tilde{X}_t - \tilde{X}_s|}{Kd(s,t)} > u\right\} \le 2 \exp\{-u^2/2\}$$

for some constant K. Then by the comments following (2.3.2) we see that we can use Theorem 2.3.1 to show that $\tilde{X}_t$ is continuous a.s. Therefore the Fourier series of $f(t) = \sum_{\gamma\epsilon\Gamma} |a_\gamma b_\gamma|\gamma(t)$, $t \epsilon G$ is a continuous function and so

$$f(0) = \sum_{\gamma\epsilon\Gamma} |a_\gamma b_\gamma| = \sum_j \left(\sum_{\gamma\epsilon\Gamma_j} |a_\gamma|^2\right)^{1/2} < \infty .$$

This completes the proof of the theorem.

REMARK 1.8. Using the results of Chapter VI it is easy to see that the converse of the preceding theorem also holds. That is, any partition which verifies the conclusion of Theorem 1.6 must be a Sidon partition. Also the reader will note that $\bigcup_{j\in J} \Gamma_j$ need not be equal to Γ.

2. *Almost sure almost periodicity*

So far we have considered random Fourier series on either compact groups or compact subsets of locally compact groups. However, the conditions which have been derived also provide a criterium for these series to be bounded continuous functions on the whole group, in which case they are almost periodic functions. This can be shown by either repeating the Dudley-Fernique argument in this setting or by compactifying the group and using results from previous chapters. We shall consider these questions in this section.

To state our results in full generality, it will be convenient to work with the following notion. Let T be an arbitrary set; we will say that a random process $\{X_t | t \in T\}$ is lattice-bounded if for each $\varepsilon > 0$ there exists an $R > 0$ such that for any finite subset $S \subset T$ we have $\mathbf{P}(\{\sup_{t\in S}|X_t| > R\}) < \varepsilon$. This means that $\{X_t | t \in T\}$ is a lattice-bounded subset of the lattice $L_0(d\mathbf{P})$.

Let G be a locally compact group. A real or complex Gaussian process $\{X(t) | t \in G\}$ will be called left stationary if, for each h in G, the distribution of the translated process $\{X(h+t) | t \in G\}$ is the same as that of original process $\{X(t) | t \in G\}$. When this holds, then the pseudo-metric ρ defined as usual for all $s, t \in G$, by

$$\rho(s,t) = (E|X(s) - X(t)|^2)^{1/2} \tag{2.1}$$

is invariant under left translation. (Actually most of the properties of left stationary Gaussian process which are considered in this book are shared by all the Gaussian processes for which the associated pseudo-metric ρ is left translation invariant.) Now let $\{X(t) | t \in G\}$ be a left stationary Gaussian process, let ρ be as above and let K be a compact subset of

G with non-empty interior. The reader who is familiar with the proofs of the Dudley-Fernique theorem (with the modifications indicated in Chapter II, §3) will easily check that $\{X(t) | t \in K\}$ is lattice-bounded if and only if the entropy condition, $\int_0^1 (\mathrm{Log}\, N_\rho(K, \varepsilon))^{1/2} d\varepsilon < \infty$, is satisfied. Moreover, we know that this happens if and only if $\{X(t) | t \in K\}$ has a version with continuous sample paths (see Theorem 2.3.1). In other words, once we have made the choice to work on the left side only, there is no difficulty to extend Theorem 2.3.2 and its converse to the non-Abelian case. It is a natural question to ask: When is $\{X(t) | t \in G\}$ lattice-bounded? The answer is given in the next theorem which can be proved easily by repeating the arguments of the Dudley-Fernique theorem given in [9].

THEOREM 2.1. *A real or complex left stationary Gaussian process* $\{X(t) | t \in G\}$ *is lattice-bounded if and only if*

$$\int_0^1 (\mathrm{Log}\, N_\rho(G, \varepsilon))^{1/2} d\varepsilon < \infty . \tag{2.2}$$

The main point of this section is that the preceding theorem can be considered as included in the compact case by introducing the Bohr compactification $\overline{G}$ of G. (For the definition of $\overline{G}$, see [17], §26.11 in the Abelian case and [5], §16.1 and 16.2 in the non-Abelian case.) Indeed, when the global entropy condition (2.2) is satisfied, it is possible to show that, up to a change of version, the function $t \to X(t)$ is almost surely an almost periodic function on G, and therefore it extends a.s. to a continuous function on $\overline{G}$. The a.s. almost periodicity can be shown directly but we prefer to adopt a more general point of view. We denote by $\overline{m}$ the normalized Haar measure on $\overline{G}$. Although in the non-Abelian case, the embedding $G \to \overline{G}$ is not necessarily injective, this does not affect the argument below and we still call, albeit abusively, $\overline{G}$ the Bohr compactification of G. Assume that $N_\rho(G, \varepsilon)$ is finite for each $\varepsilon > 0$ (in particular, this is true if (2.2) holds). Consider the left stationary Gaussian process $X : G \to L_2(\Omega, \mathcal{F}, P)$. For any $\varepsilon > 0$ we can find $t_1, \cdots, t_N$ in G such that for any $x \in G$ there is a j such that

$$\rho(t_j, x) = \sup_{t \in G} \|X(t+x) - X(t+t_j)\|_{L_2} < \varepsilon .$$

It follows from this that the Hilbert space valued function X is almost periodic. Therefore (cf. [5] Theorem 16.21 which obviously extends with the same proof to the case of Banach space valued functions) the function X extends uniquely to a continuous function $X^- : \overline{G} \to L_2(\Omega, \mathcal{F}, P)$. Let $\overline{\rho}$ be the pseudo-metric associated with X^- as in (2.1). It is easy to check that $\{X^-(t) | t \in G\}$ is again a left stationary Gaussian process which is lattice-bounded if and only if $\{X(t) | t \in G\}$ is also lattice-bounded. Since obviously $N_\rho(G, \varepsilon) = N_{\overline{\rho}}(\overline{G}, \varepsilon)$ for each $\varepsilon > 0$, we can immediately deduce Theorem 2.1 from Theorem 5.3.4. Moreover, if (2.2) is satisfied, we know that the process $\{X^-(t) | t \in \overline{G}\}$ must have a version with continuous sample paths. Therefore, we have, a fortiori

$$E \int_{\overline{G}} |X^-(t)|^2 \, \overline{m}(dt) < \infty . \tag{2.3}$$

This implies that the original process $\{X(t) | t \in G\}$ has a discrete spectral structure. Indeed, assume for simplicity that G is Abelian and consider the case of a complex valued stationary process $\{X(t) | t \in G\}$. Then (2.3) clearly implies that X^- (and therefore X) admits an expansion as a series

$$X(t) = \sum_{n=1}^{\infty} a_n \tilde{g}_n \gamma_n(t)$$

where $\{a_n\}$ is a sequence of complex numbers, $\{\gamma_n\}$ is a sequence of continuous characters on G and $\{\tilde{g}_n\}$ is an independent identically distributed sequence of standard complex valued Gaussian random variables. Of course, there is an analogous expansion of the form described in (2.3.18) in the real case. Moreover, a similar expansion exists also in the non-Abelian case, we leave this to the reader.

Now that the Gaussian case has been worked out, one can easily obtain, as a corollary of Theorem 5.3.5, the following result which for simplicity we state in the Abelian case only.

THEOREM 2.2. *Consider a locally compact Abelian group* G. *Let* $\{\gamma_n\}$ *be a sequence of continuous characters on* G *and let* $\{a_n\}$ *be a sequence of complex numbers such that*

$$\sum |a_n|^2 < \infty .$$

We consider again a Bernoulli sequence $\{\varepsilon_n\}$ *and a sequence of complex valued random variables* $\{\xi_n\}$ *independent of* $\{\varepsilon_n\}$. *We consider the process*

$$(2.4) \qquad Z(t) = \sum_{1}^{\infty} a_n \varepsilon_n \xi_n \gamma_n(t), \qquad t \in G .$$

We assume as before that

$$\inf E|\xi_n| > 0 \quad \textit{and} \quad \sup_n E|\xi_n|^2 < \infty .$$

Then the following properties are equivalent:

(i) $\{Z(t) | t \in G\}$ *is lattice bounded.*

(ii) *The series (2.4) converges a.s. in the Banach space* $C_b(G)$ *of continuous bounded functions on* G, *equipped with the* sup *norm.*

(iii) $\int_0^1 (\operatorname{Log} N_\rho(G, \varepsilon))^{1/2} d\varepsilon < \infty$ *where* ρ *is defined by*

$$\rho(s,t) = \left(\sum_{1}^{\infty} |a_n|^2 |\gamma_n(s) - \gamma_n(t)|^2\right)^{1/2}, \quad s, t \in G .$$

(iv) $\int_0^\infty \left(\operatorname{Log} \frac{1}{M(\varepsilon)}\right)^{1/2} d\varepsilon < \infty$ *where*

$$M(\varepsilon) = \overline{m}(\{x \in \overline{G} | \overline{\rho}(0, x) < \varepsilon\}) .$$

The reader will remark that (ii) implies that $t \to \lim_{n\to\infty} \sum_1^n a_k \varepsilon_k \xi_k \gamma_k(t)$ is almost surely an almost periodic function on G.

3. *On left and right almost sure continuity*

In this section, we discuss the connection between the two spaces $C_{a.s.}(G)$ which arise respectively when we work with left or right translations on a non-Abelian compact group G. We keep the same notation as in Chapter VI. In Chapter VI, we consistently defined the space $C_{a.s.}(G)$ using translation on the left. Clearly, we can define similarly the space $C^r_{a.s.}(G)$ using right instead of left translation. To be more specific, we define, $C^r_{a.s.}(G)$ as the space of all functions f in $L_2(G)$ such that the random Fourier series

$$\sum_{i\in\Sigma} d_i \operatorname{tr} U_i \tilde{G}_i \hat{f}(i)$$

is almost surely continuous on G. We equip the space $C^r_{a.s.}(G)$ with the norm

$$[[f]]_r = E\|\sum_{i\in\Sigma} d_i \operatorname{tr} U_i \tilde{G}_i \hat{f}(i)\|_{C(G)} .$$

It is easy to see that if $\hat{f}(i)$ commutes with $\hat{f}(i)^*$ for each $i \in \Sigma$, then $[[f]] = [[f]]_r$. However, in general, $C_{a.s.}(G)$ does not coincide with $C^r_{a.s.}(G)$, as the following simple example shows.

EXAMPLE 3.1. There is a compact group G such that $C_{a.s.}(G) \neq C^r_{a.s.}(G)$.

Proof. We use a sequence of finite groups $G(n)$ defined as follows. For each integer n, $G(n)$ is the group of all $n\times n$ matrices $(a_{ij})_{i\le i,j\le n}$ which can be written in the following way: there exists a choice of signs $\varepsilon_i = \pm 1$, $i = 1,2,\cdots,n$ and a permutation π of $\{1,2,\cdots,n\}$ such that $a_{ij} = \varepsilon_i$ if $j = \pi(i)$ and $a_{ij} = 0$ otherwise. Clearly, $G(n)$ is a finite subgroup of the unitary group $U(n)$; moreover the representation denoted

$U_n : G(n) \to B(\ell_2^n)$, which associates to each matrix in $G(n)$ the corresponding unitary operator on ℓ_2^n, is irreducible (cf. [17], §29.47). For any $n \times n$ matrix $b = (b_{ij})_{1 \le i,j \le n}$, it is easy to see that

$$\sup_{a \in G(n)} |\mathrm{tr}\, U_n(a)\, b| = \sup_{\pi} \sum_{i=1}^{n} |b_{\pi(i)i}| . \tag{3.1}$$

Consider the function $f : G(n) \to \mathbf{R}$ defined for all $a \in G(n)$ by $f(a) = \sum_{j=1}^{n} a_{1j}$. Using (3.1), one can check immediately that

$$[[f]] = 1/\sqrt{n}\, E\Big(\sum_{j=1}^{n} |\tilde{g}_j|\Big) ,$$

$$[[f]]_r = E \sup_{1 \le j \le n} |\tilde{g}_j| ,$$

where $\{\tilde{g}_j\}$ is a sequence of independent complex valued Gaussian random variables with mean zero and variance 1. In particular, as n tends to infinity, we have $[[f]] \sim \sqrt{n}$ while $[[f]]_r \sim \sqrt{\mathrm{Log}\, n}$. Therefore, if we consider $G = \prod_{n=1}^{\infty} G(n)$, we have $C_{a.s.}(G) \ne C^r_{a.s.}(G)$ since, by the preceding computation, these two Banach spaces have non-equivalent norms.

Obviously, if G is Abelian, we have $C_{a.s.}(G) = C^r_{a.s.}(G)$ with equality of the corresponding norms. More generally, if the dimensions $\{d_i\}_{i \in \Sigma}$ of the irreducible representations on G are uniformly bounded, then it is not difficult to check that $C_{a.s.}(G) = C^r_{a.s.}(G)$ and the corresponding norms are equivalent. We conjecture that the converse also holds, i.e. that $C_{a.s.}(G) = C^r_{a.s.}(G)$ implies that $\sup_{i \in \Sigma} d_i < \infty$. This conjecture is suggested by an analogous conjecture of C. S. Herz concerning the spaces of bounded multipliers on $L_p(G)$ $p \ne 2$; indeed, the results of

Chapter VI show that $C_{a.s.}(G) = C^r_{a.s.}(G)$ if and only if a similar identity holds for the spaces of bounded multipliers from $L_2(G)$ into $L_\psi(G)$ on the left and on the right.

4. *Generalizations*

Consider a random Fourier series on a locally compact Abelian group as given in (1.1.11) but remove condition (1.1.10) so that $\{\xi_\gamma\}$ is simply a sequence of complex valued random variables independent of $\{\varepsilon_\gamma\}$. Using the notation provided at the beginning of Chapter III we have that this series converges uniformly a.s. if and only if

$$I(\sigma(u,\omega_1)) < \infty \text{ a.s. with respect to } (\Omega_1, \mathcal{F}_1, P_1) \tag{4.1}$$

where

$$\sigma(u,\omega_1) = \Big(\sum_{\gamma\epsilon A} |a_\gamma|^2 |\xi_\gamma(\omega_1)|^2 |\gamma(u)-1|^2\Big)^{1/2}.$$

(Note that $\sum_{\gamma\epsilon A} |a_\gamma|^2|\xi_\gamma(\omega_1)|^2 < \infty$ a.s. is an obvious necessary condition.) When $\{\xi_\gamma\}$ satisfies (1.1.10) we can express (4.1) deterministically, solely as a function of $\{\gamma|\gamma\epsilon A\}$ and $\{a_\gamma\}$. We would like to obtain similar deterministic statements when $\{\xi_\gamma\}$ satisfies different conditions than (1.1.10). Let E_1 denote expectation with respect to $(\Omega_1, \mathcal{F}_1, P_1)$. Then, since

$$E_1\{I(\sigma(u,\omega_1))\} \le I(E_1\sigma(u,\omega_1)), \tag{4.2}$$

the right side of (4.2) finite provides a sufficient condition for $I(\sigma(u,\omega_1))$ to be finite a.s. with respect to $(\Omega_1, \mathcal{F}_1, P_1)$. This approach was used to obtain Theorem 2.9 [39], which although given for R^1 extends immediately to the cases considered here. However, the sufficient condition (2.42) of Theorem 2.9 [39] is not necessary.

Cuzick and Lai make a useful observation in Theorem 3 [4]. Let $\{b_\gamma\}$ be a sequence of real numbers such that

(4.3) $$\sum_{\gamma \epsilon A} E\{\min(|a_\gamma \xi_\gamma|, 1) I_{[|\xi_\gamma|>b_\gamma]}\} < \infty .$$

Then, a sufficient condition for the uniform convergence a.s. of (1.1.11) is that

(4.4) $$I\left(E\left(\sum_{\gamma \epsilon A} |a_\gamma|^2 |\xi_\gamma|^2 I_{[|\xi_\gamma|\leq b_\gamma]} |\gamma(u)-1|^2\right)^{1/2}\right) < \infty .$$

This result is obtained as follows: By standard truncation arguments one can show that (4.3) implies that (1.1.11) converges uniformly a.s. if and only if

(4.5) $$\sum_{\gamma \epsilon A} a_\gamma \varepsilon_\gamma \xi_\gamma I_{[|\xi_\gamma|\leq b_\gamma]} \gamma(x), \qquad x \epsilon K$$

converges uniformly a.s. Thus we use (4.2) applied to (4.5) to obtain (4.4) as in the proof of Theorem 1.1.1. The condition given in (4.4) can be checked in specific cases and generally leads to sharper results than (4.2) alone. (Additional results concerning the uniform convergence a.s. of random Fourier series when the random variables do not have finite second moments can also be found in [4].)

We have generalized the sufficiency part of Theorem 1.1.1 as follows: Let $X(t, \omega, \omega_1)$, $t \epsilon T$, T a metrizable space, be a complex valued stochastic process defined on the product probability space $(\Omega \times \Omega_1, \mathcal{F} \times \mathcal{F}_1, P \times P_1)$ satisfying, for $q \geq 2$,

(4.6) $$E_\omega\left[\exp\left|\frac{X(t,\omega,\omega_1) - X(s,\omega,\omega_1)}{kd(s,t,\omega_1)}\right|^q\right] \leq 2$$

where k is some constant and, for each $\omega_1 \epsilon \Omega_1$, $d(s,t,\omega_1)$ is a norm or pseudo-norm on T with the property of $\| \ \|_\Omega$ defined just before Lemma 2.2.3. Also assume that $d(s,t,\omega_1)$ is translation invariant a.s. with respect to $(\Omega_1, \mathcal{F}_1, P_1)$ and that $Ed(s,t,\omega_1) \leq \sigma(s-t)$. Then

(4.7) $$\int_0^\infty (\log N_\sigma(T,\varepsilon))^{1/q} d\varepsilon < \infty$$

is a sufficient condition for $X(t, \omega, \omega_1)$ to have a version with continuous sample paths. As an application of this result let $\{\theta_k\}$ be real valued independent identically distributed stable random variables of index p, $1 < p < 2$, i.e.

$$E[e^{iv\theta_1}] = e^{-|v|^p}, \quad -\infty < v < \infty .$$

Consider, as a specific case of (1.1.11),

$$\sum_{\gamma \epsilon A} a_\gamma \theta_\gamma \gamma , \tag{4.8}$$

and define

$$\sigma_p(u) = \Big(\sum_{\gamma \epsilon A} |a_\gamma|^p |\gamma(u) - 1|^p\Big)^{1/p} . \tag{4.9}$$

Then

$$I_p(\sigma_p) = \int_0^{\mu_2} \frac{\overline{\sigma}_p(s)}{s\left(\log \frac{4\mu_4}{s}\right)^{1/p}} ds < \infty \tag{4.10}$$

is a sufficient condition for the uniform convergence a.s. of the series (4.9). This result follows from the one mentioned in (4.7) when $p = q/(q-1)$. It is possible that (4.10) is also necessary for the uniform convergence a.s. of (4.8). (Note that the result mentioned in (4.6) holds more generally for convex functions other than $(\ )^q$ with similar extensions of the form (4.7) and (4.10). These questions are still being studied.)

The series (1.4) with $\{\xi_k\}$ real valued random variables is a special case of the stochastic integral

$$\int_{-\infty}^{\infty} e^{i\lambda t} \xi(d\lambda, \omega), \qquad t \,\epsilon\, [0, 1] , \tag{4.11}$$

where $\xi(d\lambda, \omega)$ is a random measure with sign-invariant increments. Fernique [10] obtained an analogue of the sufficient part of Theorem 1.1.1 in the case

$$E\left|\int_{\lambda_1}^{\lambda_2} \xi(d\lambda, \omega)\right|^2 = F(\lambda_2) - F(\lambda_1)$$

where F is a distribution function on R, (and similarly for R^n). In fact, his result is that the process given in (4.11) has a version with continuous sample paths if $I(\sigma) < \infty$, where, in this case,

$$\sigma(u) = \left(\int_{-\infty}^{\infty} \sin^2 \frac{\lambda u}{2}\, dF(\lambda)\right)^{1/2} .$$

The case in which $\xi(d\lambda, \omega)$ is a randomly scattered independent stable measure is considered in [1] and related questions are considered in [15]. In these cases analogues of (4.10) can be obtained.

Another generalization is to consider series of the form (1.4) in which $\{\lambda_k\}$ are real valued random variables independent of $\{\varepsilon_k\}$ and $\{\xi_k\}$. This is a meaningful generalization because it includes the empirical characteristic function which is of interest to statisticians. Most of our results extend to this case since it only requires one more conditional expectation. Work along this line can be found in [37] with further applications in [38].

Most of the remarks of this section have appropriate realizations for random Fourier series on compact non-Abelian groups.

It appears that the most difficult open question is to find *necessary* conditions for the uniform convergence a.s. of the series (1.1.11) when $\{\xi_\gamma\}$ does not satisfy (1.1.10). In our work presented here, in which $\{\xi_\gamma\}$ does satisfy (1.1.10), we make critical use of Fernique's necessary condition for the continuity of stationary Gaussian processes. This result in turn depends on Slepian's lemma or its equivalent and we do not have a version of Slepian's lemma for non-Gaussian processes.

REFERENCES

[1] Araujo, A. and M. B. Marcus, Stable processes with continuous sample paths, Proc. Second International Conference on Probability in Banach Spaces (1978), Lecture Notes in Math. 709 (1979), 9-32, Springer-Verlag, New York.

[2] Belyaev, Yu. K., Continuity and Hölder's conditions for sample functions of stationary Gaussian processes, Proc. Fourth Berkeley Symp. Math. Statist. Prob. *2* (1961), 22-33.

[3] Billingsley, P., *Convergence of Probability measures*, (1968), Wiley, New York.

[4] Cuzick, J. and T. L. Lai, On random Fourier series, Trans. Amer. Math. Soc., 261 (1980), 58-80.

[5] Dixmier, J., *Les C^* algèbres et leurs représentations*, (1969), Gauthier Villars, Paris.

[6] Dudley, R. M., The sizes of compact subsets of Hilbert space and continuity of Gaussian processes, J. Funct. Anal. *1* (1967), 290-330.

[7] Edwards, R. and K. Ross, On p-Sidon sets. J. Functional Anal., 15 (1974), 404-427.

[8] Fernique, X., Des résultats nouveaux sur les processus gaussien, C. R. Acad. Sci. Paris Ser. A *278* (1974), 363-365.

[9] ________, Régularité des trajectoires des fonctions aléatoires gaussiennes, Lecture Notes in Mathematics, 480 (1975), 1-96.

[10] ________, Continuité et théorème central limite pour les transformées de Fourier des mesures aléatoires du second ordre, Z. Warscheinlichketisth., *42* (1978), 57-66.

[11] Figa-Talamanca, A. and D. Rider, A theorem of Littlewood and lacunary series for compact groups, Pacific J. Math., 16 (1966), 505-514.

[12] ________, A theorem on random Fourier series on non-commutative groups, Pacific J. Math. 21 (1967), 487-492.

[13] Figa-Talamanca, A., Random Fourier series on compact groups, Theory of group representations and Fourier analysis, C.I.M.E. (1970), Rome Ed. Cremonese, 1971.

[14] Garsia, A., Combinatorial inequalities and smoothness of functions, Bull. Amer. Math. Soc. *82* (1976), 157-170.

[15] Giné, E. and M. B. Marcus, Some results on the domain of attraction of stable measures on C(K), Probability and Mathematical Statistics, (1981).

[16] Hardy, G. H., Littlewood, J. E. and Pólya, G., *Inequalities* (1934), Cambridge Univ. Press, England.

[17] Hewitt, E. and K. Ross, *Abstract Harmonic Analysis*, Vol. I and Vol. II. (1970), Springer Verlag, Berlin.

[18] Hoffmann-Jørgensen, J., Sums of independent Banach space valued random variables, Studia Math., *52* (1974), 159-186.

[19] ————, Probability in B-spaces, Ecole d'Eté de Probabilitiés de Saint-Flour, VI-1976, Lecture Notes in Math., 598, Springer-Verlag, Berlin.

[20] Hunt, G. A., Random Fourier transforms, Trans. Amer. Math. Soc. 71 (1951), 38-69.

[21] Ito, K. and Nisio, M., On the convergence of sums of independent Banach space valued random variables, Osaka Math. J. *5* (1968), 35-48.

[22] Jain, N., Central limit theorems and related questions in Banach space, Proc. AMS Probability Symp., Urbana, Illinois (1976).

[23] Jain, N. C. and Marcus, M. B., Sufficient conditions for the continuity of stationary Gaussian processes and applications to random series of functions, Ann. Inst. Fourier (Grenoble) *24* (1974), 117-141.

[24] ————, Integrability of infinite sums of independent vector-valued random variables, Trans. Amer. Math. Soc. *212* (1975), 1-36.

[25] ————, Continuity of sub-gaussian processes, *Advances in Probability*, Vol. 4, (1978), M. Dekker, New York.

[26] Kahane, J. P., *Some random series of functions*, (1968), D. C. Heath, Lexington, Mass., U.S.A.

[27] Krasnoselskii, M. A. and Rutickii, Y., *Convex functions and Orlicz spaces*, (1961), Noordhof, Groningen.

[28] Kwapień, S., A theorem on the Rademacher series with vector coefficients, Probability in Banach spaces, Lecture Notes in Math. 526, 157-158, Springer-Verlag, Berlin.

[29] Larsen, R., *An Introduction to the Theory of Multipliers*, (1971), Springer-Verlag, Berlin.

[30] Lindenstrauss, J. and A. Pełczyński, Absolutely summing operators in $\mathcal{L}_p$ spaces and their applications, Studia Math. 29 (1968), 275-326.

[31] Lopez, J. and K. Ross, Sidon sets, Lecture notes in pure and applied math. No. 13, (1975), Marcel Dekker, New York.

[32] Malliavin, M. P. and P. Malliavin, Caractérisation arithmétique d'une classe d'ensembles de Helson, C. R. Acad. Sc. Paris, A, *164*(1967), 192-193.

[33] Marcus, M. B., A comparison of continuity conditions for Gaussian processes, Ann. of Prob., 1 (1973), 123-130.

[34] ________, Continuity of Gaussian processes and Random Fourier series, Ann. of Probability, *1* (1973), 968-981.

[35] ________, Uniform convergence of random Fourier series, Ark. Mat. *13* (1975), 107-122.

[36] ________, Continuity and the central limit theorem for random trigonometric series, Z. Warhscheinlichkeitsch., *42* (1978), 35-56.

[37] ________, Weak convergence of the empirical characteristic function, Ann. of Prob., (1981).

[38] Marcus, M.B. and W. Phillip, Almost sure invariance principles for C(K) valued random variables with applications to random Fourier series and the empirical characteristic process, Trans. Amer. Math. Soc., (1981).

[39] Marcus, M.B. and Pisier, G., Necessary and sufficient conditions for the uniform convergence of random trigonometric series, Lecture Note Series, No. 50, (1978), Arhus University, Arhus, Denmark.

[40] ________, Random Fourier series on locally compact Abelian groups, Seminaire de Probabilities XIII, Université de Strasbourg 1977/1978, Lecture Notes in Math. 709 (1979), 72-89, Springer-Verlag, New York.

[41] Maurey, B., Type et cotype dans les espaces de Banach munis de structure locale inconditionnelle, Exposé 24-25, Séminaire Maurey-Schwartz 73/74, Ecole Polytechnique, Paris.

[42] Maurey, B. and G. Pisier, Séries de variables aléatoires vectorielles independantes et propriétés géométriques des espaces de Banach, Studia Math. 58 (1976), 45-90.

[43] Neveu, J., *Discrete parameter martingales*, (1975), North Holland, Amsterdam.

[44] Paley, R.E.A.C. and Zygmund, A., On some series of functions (1) (2) (3), Proceedings of Cambridge Phil. Soc., *26*(1930), 337-357, *26* (1930), 458-474, *28*(1932), 190-205.

[45] Pisier, G., Sur le theoreme limite central et la loi du logarithme itéré dans les espaces de Banach, Séminaire Maurey-Schwartz 1975-76, Ecole Polytechnique, Paris.

[46] ________, Les inégalités de Khintchine-Kahane d'aprés C. Borell, Séminaire sur la géométrie des espace de Banach, 1977-78, Ecole Polytechnique, Palaiseau.

[47] Pisier, G., Sur l'espace de Banach des séries de Fourier aléatoires presque sûrement continues, Exposé No. 17-18, Séminaire sur la géométrie des espace de Banach, 1977-78, Ecole Polytechnique, Palaiseau.

[48] ———, Ensembles de Sidon et processus gaussiens, C. R. Acad. Sci. Paris A 286 (1978), 671-674.

[49] ———, A remarkable homogeneous Banach algebra, Israel J. Math., 34 (1979), 38-44.

[50] Preston, C., Banach spaces arising from some integral inequalities, Indiana U. Math. J. 20 (1971), 997-1015.

[51] Rider, D., Randomly continuous functions and Sidon sets, Duke Math. J. 42 (1975), 759-764.

[52] ———, Random Fourier series, Symposia Math., 22 (1977), 93-106.

[53] Rudin, W., Trigonometric series with gaps, J. Math. and Mech. 9 (1960), 203-227.

[54] ———, *Fourier Analysis on groups*, (1962), Interscience, New York.

[55] Salem, R. and Zygmund, A., Some properties of trigonometric series whose terms have random signs, Acta Math. 91 (1954), 245-301.

[56] Tomczak-Jaegermann, N., On the moduli of smoothness and convexity and the Rademacher averages of trace classes S_p $(1 \le p < \infty)$, Studia Math. 50 (1974), 163-182.

[57] Vilenkin, N. Ja., Special functions and the theory of group representations, (1968), Amer. Math. Soc. Transl. V. 22, Providence.

[58] Pisier, G., De nouvelles caractérisations des ensembles de Sidon. Advances in Maths., Supplementary Studies, (1981), (L. Nachbin, Editor), to appear.

INDEX OF TERMINOLOGY

INDEX OF NOTATIONS

Library of Congress Cataloging in Publication Data

Marcus, Michael B.
Random Fourier series with applications to harmonic analysis.

Bibliography: p.
1. Fourier series. 2. Harmonic analysis. I. Pisier, Gilles, 1950- . II. Title.
QA404.M32 1981 515'.2433 81-47145
ISBN 0-691-08289-8 AACR2
ISBN 0-691-08292-8 (pbk.)